Human Anatomy & Physiology

Part II

David G. Gantt, Ph.D.
Georgia Southern University

Denson K. McLain, Ph.D.
Georgia Southern University

Ann E. Pratt, Ph.D.
Georgia Southern University

AudioText

Other AudioText publications:

Biology-On-Tape: *Understanding the Cell*
Biology-On-Tape: *Genetics, Diversity, and the Biosphere*
Biology-On-Tape: *Human Anatomy & Physiology - Part I*
Biology-On-Tape: *Human Anatomy & Physiology - Part II*

Understanding the Cell
Genetics, Diversity, and the Biosphere
Human Anatomy & Physiology - Part I

AudioText, Inc.
P.O. Box 690791
Houston, TX 77269-0791

International Standard Book Number:
1-884612-07-5

Library of Congress Catalog Card Number:
95-83836

123456789 - 99 98 97

Acknowledgements

We owe special thanks to John Eylers and Michael Buono for their many suggestions and contributions to this text.

TABLE OF CONTENTS

Chapter 1   The Endocrine System   1

Chapter 2   The Blood   12

Chapter 3   The Heart   22

Chapter 4   The Circulatory System   31

Chapter 5   The Lymphatic & Defense Systems   50

Chapter 6   The Respiratory System   60

Chapter 7   The Urinary System   71

Chapter 8   The Digestive System   84

Chapter 9   The Reproductive System   98

INDEX   112

Chapter 1

THE ENDOCRINE SYSTEM

The **endocrine system** is composed of glands that regulate physiological activities through the secretion of hormones. The nervous system controls endocrine secretions that produce long-duration, generalized alterations of metabolic activity affecting homeostatic controls, embryological development, growth, sexual maturation, and reproduction. **Endocrine glands**, unlike exocrine glands, do not have ducts to carry their products. Rather, each hormone is released directly into the blood stream which carries it to target cells that have specific cell-surface receptors for it. Target cells may be assembled in target tissues such as a single gland or organ or they may be scattered throughout the body.

Once a peptide hormone binds to a plasma membrane receptor the production of an intracellular "second messenger" such as cyclic AMP may be initiated. The second messenger functions as an enzyme inhibitor or activator, thereby altering cellular metabolic activity. Other peptide hormones change plasma membrane permeability or membrane potential by opening or closing ion channels. Steroid hormones enter the cell and bind directly to DNA, affecting the transcription of messenger RNAs that themselves direct the synthesis of new enzymes.

Peptide hormones are amino acid-based. These are either biogenic amines, which are derivatives of various amino acids, or peptides, which are chains of amino acids. Derivatives of the amino acid, tyrosine, are the thyroid hormones, catecholamines, epinephrine, and norepinephrine. The amino acid, histidine, is converted into histamine by mast cells in connective tissues. And, the amino acid, tryptophan, is converted into the hormones serotonin and melatonin. Peptides that function as hormones vary in length from a few amino acids (oligopeptides) to a few hundred amino acids (polypeptides). Polypeptide, or protein, hormones are too large to diffuse through the plasma membrane and, therefore, react with receptors on the cell surface.

Many hormones, the **steroid hormones**, are derivatives of the chemical, cholesterol. Examples are the adrenocortical hormones and sex hormones such as estrogen and testosterone. Steroid hormones, and the small thyroid hormones, are lipid-soluble. Thus, they can cross the plasma membrane to bind to receptors within target cells.

Other hormones, the **eicosanoids**, are chemical derivatives of the fatty acid, arachidonic acid. Examples are prostaglandins and leukotrenes. They differ from other hormones in having only localized effects.

The hypothalamus of the brain regulates both the nervous and endocrine systems. It exerts direct neural control on the medulla of the adrenal gland and regulates hormonal secretions of the pituitary gland via its own endocrine secretions.

The pea-sized, two-lobed **pituitary gland**, or hypophysis, lies at the base of the brain in a bony depression, the sella turcica of the sphenoid bone. The pituitary is attached to the hypothalamus by a stalk, the **infundibulum**. Regulating-hypothalamic hormones are released into surrounding interstitial fluids from which they diffuse into a capillary network, the primary plexus. The capillaries are derived from branching of the superior hypophyseal arteries that are themselves branches of the internal carotid arteries. These relatively permeable capillaries are connected to capillaries at the anterior lobe of the pituitary by hypophyseal portal veins that branch among target endocrine cells.

This circulatory arrangement is unique because the vessels between the median eminence of the hypothalamus and the anterior pituitary carry blood from one capillary network to another. The circulation is one-way because any chemical released by downstream cells must pass through the circulatory system before reaching upstream capillaries at the start of the portal system.

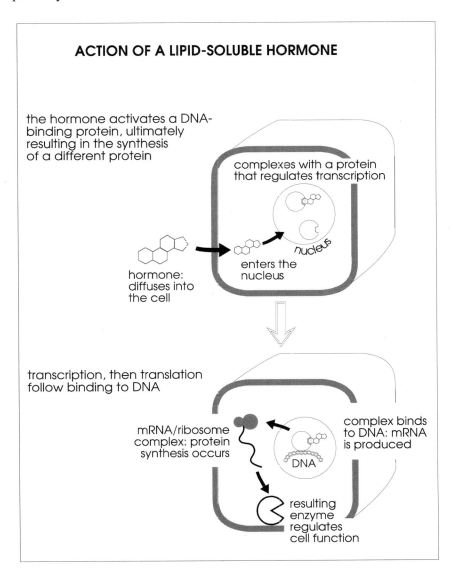

ACTION OF A LIPID-SOLUBLE HORMONE

the hormone activates a DNA-binding protein, ultimately resulting in the synthesis of a different protein

complexes with a protein that regulates transcription

enters the nucleus

nucleus

hormone: diffuses into the cell

transcription, then translation follow binding to DNA

mRNA/ribosome complex: protein synthesis occurs

complex binds to DNA: mRNA is produced

DNA

resulting enzyme regulates cell function

There are five principal types of **anterior pituitary gland cells** that secrete seven major hormones. **Gonadotrophs** are cells that produce the gonadotrophins, follicle stimulating hormone (FSH) and luteinizing hormone (LH), that regulate gonadal activity. In females, follicle stimulating hormone promotes maturation of oocytes in the ovaries and stimulates the release of estrogen by ovarian follicle cells. In males, FSH stimulates sperm production in the testes. Luteinizing hormone promotes ovarian secretion of progesterone and estrogen and induces ovulation. In males, LH stimulates the production of male sex hormones, primarily testosterone, by interstitial cells of the testes.

Somatotrophs are pituitary cells that produce human growth hormone, or somatotropin. Somatotropin promotes growth by indirectly stimulating protein synthesis. That is, liver cells respond to somatotropin by releasing insulin-like somatomedins that bind to receptor sites on the surfaces of a variety of cells. Somatomedins increase the rate of amino acid transport into cells where they are used in protein synthesis. The secretion of somatotropin is regulated by two hypothalamic hormones. Growth releasing hormone, or somatocrinin, promotes secretion of somatotropin while growth inhibiting hormone, or somatostatin, inhibits the secretion.

Thyrotrophs are anterior pituitary cells that synthesize thyroid-stimulating hormone, controlling the secretions of the thyroid gland. **Lactotrophs** are cells that synthesize prolactin, which, in conjunction with other hormones, stimulates glandular tissue development in female breasts during pregnancy. Prolactin also stimulates milk production. **Corticotrophs** are cells that synthesize adrenocorticotrophic hormone, which stimulates the release of glucocorticoids by the cortex of the adrenal gland. Corticotrophs also secrete melanocyte-stimulating hormone, promoting melanin pigment production by skin melanocytes. Melanin provides protection from UV radiation.

The **posterior lobe of the pituitary** does not synthesize hormones but does store and release antidiuretic hormone, or vasopressin, and oxytocin. **Antidiuretic hormone** is released in response to changes in blood volume and regulates water balance by causing reabsorption of water in the kidneys and by promoting perspiration. **Oxytocin** stimulates smooth muscle contraction in the walls of the uterus during childbirth. Oxytocin also causes milk secretion. Both antidiuretic hormone and oxytocin are synthesized by neurosecretory cells of the hypothalamus. They then travel along axons to the posterior pituitary lobe.

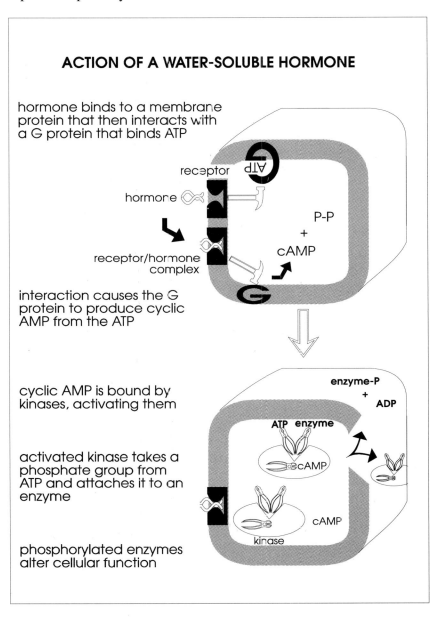

ACTION OF A WATER-SOLUBLE HORMONE

hormone binds to a membrane protein that then interacts with a G protein that binds ATP

interaction causes the G protein to produce cyclic AMP from the ATP

cyclic AMP is bound by kinases, activating them

activated kinase takes a phosphate group from ATP and attaches it to an enzyme

phosphorylated enzymes alter cellular function

The **thyroid gland** consists of two lobes of follicle cells that wrap around the anterior surface of the trachea, just below the larynx. When low blood levels of the thyroid hormones, T_3 and T_4, are detected by the hypothalamus, **thyrotropin releasing hormone** is secreted. Thyrotropin releasing hormone is carried by hypophyseal portal veins to the anterior pituitary gland, where it stimulates the secretion of **thyroid-stimulating hormone**. Thyroid-stimulating hormone causes follicle cells to release thyroid hormones that elevate the metabolic rate. Elevated levels of T_3 and T_4 then inhibit the release of thyrotropin releasing hormone and thyroid-stimulating hormone, providing negative feedback regulation of the metabolic rate.

Thyroid hormones increase basal metabolic rate by accelerating cellular aerobic respiration and production of integral membrane proteins of the sodium-potassium ion pump. T_4 produces a gradual, long-term response as it is converted into T_3 in tissues into which it diffuses. T_3 produces a short-lived, but immediate response.

T_3, or triiodothyronine, and T_4, or tetraiodothyronine, are iodine-substituted derivatives of the amino acid, tyrosine. Goiter arises from a dietary iodine deficiency and is associated with an enlarged thyroid gland.

Parafollicular , or C, cells are endocrine cells of the thyroid that lie between follicle cells and their basement membrane. **Parafollicular cells** produce the hormone, calcitonin. **Calcitonin** regulates calcium and phosphate ion concentrations in blood by stimulating excretion by kidneys and by reducing intestinal absorption. Calcitonin also stimulates osteoblast activity and inhibits osteoclast activity, further contributing to the decrease in ion concentrations in the blood.

Two pairs of **parathyroid glands** are usually embedded in the posterior surface of the thyroid gland. Parathyroid glands produce **parathormone**, or parathyroid hormone, in response to low blood calcium levels. Parathormone elevates the circulating concentration of calcium and phosphate ions by stimulating osteoclasts to digest bone matrix while inhibiting bone building by osteoblasts, and by promoting intestinal absorption of calcium and phosphate ions. In the kidneys, parathormone reduces urinary excretion of calcium and stimulates the synthesis of calcitriol from vitamin D which is necessary for intestinal absorption of dietary calcium.

The paired **adrenal glands** are located in fat layers, one near the top of each kidney. They consist of an outer adrenal cortex and an inner adrenal medulla that develop from different embryonic tissues and secrete different hormones. The hypothalamus controls the adrenal glands in two ways. First, the secretion of **adrenocorticotrophic releasing hormone** into the anterior pituitary gland results in the synthesis of adrenocorticotrophic hormone which acts at the adrenal cortex. Second, there are direct neural connections from the hypothalamus to the spinal cord that impact the sympathetic nerves that run from the spinal cord to the adrenal medulla.

The **adrenal cortex** has three hormone-secreting zones. The outermost **zona glomerulosa** secretes mineralocorticoids that maintain homeostatic control over electrolyte concentrations in plasma and water balance. **Aldosterone**, accounting for over 95% of the mineralocorticoids, increases renal reabsorption of sodium, chloride, and bicarbonate ions, causing the retention of water. Aldosterone also promotes increased potassium and hydrogen ion excretion in urine. One effect of these actions is to increase blood volume and pressure.

Aldosterone secretion is stimulated by dehydration which results in the loss of sodium ions, high levels of potassium ions, and decreased blood volume and pressure. Juxtaglomerular cells of the kidney secrete the enzyme, renin, in response to lowered blood pressure. In the blood, **renin** converts angiotensinogen, a plasma protein produced by the liver, into angiotensin. **Angiotensin** is converted into angiotensin II by **angiotensin converting enzyme** in the lungs. **Angiotensin II** then stimulates aldosterone secretion by the adrenal cortex.

The **middle zona fasciculata** of the adrenal cortex secretes glucocorticoids, including **cortisol** or hydrocortisone. **Glucocorticoids** regulate metabolism and resist physiological responses to stress. **Corticotropin releasing hormone**, secreted by the hypothalamus, stimulates the secretion of adrenocorticotrophic hormone by the anterior pituitary gland. Under non-stressful conditions, glucocorticoids regulate blood sugar levels and maintain blood volume by preventing the movement of water into tissues. But, during severe stress arising from infection, hemorrhage, or trauma, glucocorticoid hormones dramatically elevate blood glucose, fatty acid, and amino acid levels by stimulating gluconeogenesis and lipid and protein catabolism.

Cortisol stimulates gluconeogenesis, the production of glucose, in the liver from lactic acid and certain amino acids. Protein catabolism primarily entails the breakdown of muscle fibers, liberating amino acids into the blood. Glucocorticoids also stimulate lipolysis, the breakdown of triglycerides that releases fatty acids from adipose tissue These processes provide the substrates for ATP synthesis in response to stress.

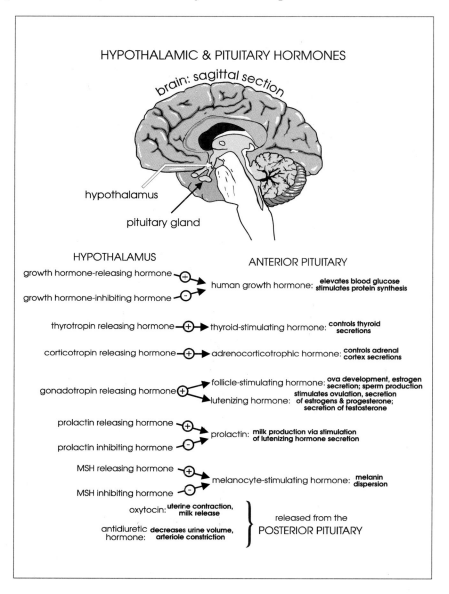

Chromaffin cells of the **adrenal medulla** synthesize and secrete epinephrine, or **adrenaline**, and norepinephrine, or **noradrenaline,** in response to stimulation by sympathetic preganglionic neurons of the autonomic nervous system. **Epinephrine** is a cardiac stimulator while **norepinephrine** causes vasoconstriction. Together, they increase the heart rate and force of contraction, increasing blood flow to skeletal muscles. Other effects are the dilation of pulmonary air passages, an increase in the rate of breathing, and a decrease in digestive activity. All of these effects increase the oxygen supply to skeletal muscles.

The **pancreas** is a long, triangular gland lying between the stomach and the duodenum of the small intestine. The pancreas functions as both an exocrine and endocrine gland. Its exocrine function is the work of acinar cells, accounting for 99% of the pancreatic mass, and entails the release of digestive enzymes into the duodenum via the pancreatic duct. Endocrine functions are performed by four types of cells clustered in the pancreatic islets, or **islets of Langerhans. Alpha cells** secrete glucagon, **beta cells** secrete insulin, **delta,** or **D, cells** secrete somatostatin, and **F cells** secrete pancreatic polypeptide.

Glucagon is secreted in response to decreased levels of blood glucose. Glucagon stimulates the synthesis of glucose from products of triglyceride, fatty acid, and protein catabolism. Glucagon also stimulates glycogenolysis, the release of glucose residues stored in glycogen in liver cells.

Insulin release is stimulated by high levels of blood glucose, as after eating, and its effect is opposite that of glucagon. Insulin lowers blood glucose levels by stimulating cellular uptake of glucose and by stimulating its conversion into glycogen in muscle, liver, and other cells. Insulin also stimulates amino acid uptake, protein synthesis, and fat storage.

Somatostatin, or growth hormone-inhibiting hormone, inhibits the secretion of insulin and glucagon and inhibits the absorption of nutrients by the intestines in response to high glucose levels.

Gonads, testes in males and ovaries in females, produce sperm and ova respectively. The gonads also function as endocrine glands, secreting hormones under the indirect control of the hypothalamus via its secretion of gonadotropin releasing hormone. Gonadotropin releasing hormone causes the anterior pituitary to secrete both follicle stimulating hormone and luteinizing hormone.

ENDOCRINE GLANDS AND THEIR HORMONES

ADRENAL CORTEX

aldosterone: **blood volume and electrolyte balance**

cortisol: **regulation of metabolism, anti-inflammatory**

ADRENAL MEDULLA

epinephrine & norepinehrine: **energy allocation to urgent contingencies**

THYROID GLAND

triiodothyronine & thyroxin: **regulates growth & development**

calcitonin: **absorption of calcium & phosphates into bone**

PARATHYROID GLAND

parathyroid hormone: **dietary absorption of calcium and magnesium, stimulates osteoclast activity, calcium reabsorption & phosphate excretion by kidneys, synthesis of calcitriol**

PANCREAS

glucagon: **elevates blood glucose levels via glycogen breakdown & glucose synthesis**

insulin: **lowers blood glucose level, stimulates synthesis of glycogen, lipids, & proteins**

somatostatin (growth hormone inhibiting hormone): **inhibits secretion of insulin & glucagon**

pancreatic polypeptide: **release of digestive enzymes by the pancreas**

OVARIES

estrogens & progesterone: **menstruation, pregnancy, lactation, & oogenesis**

relaxin: **dilates cervix prior to giving birth**

inhibin: **inhibits secretion of follicle-stimulating hormone**

TESTES

testosterone: **development of male secondary characters & spermatogenesis**

inhibin: **inhibits secretion of follicle-stimulating hormone**

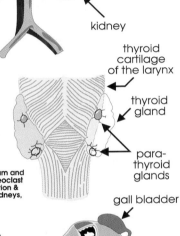

adrenal gland

kidney

thyroid cartilage of the larynx

thyroid gland

para- thyroid glands

gall bladder

small intestine

pancreas

ovaries

uterus

Luteinizing hormone stimulates cells in the testes to secrete testosterone, the primary male sex hormone, or androgen. **Testosterone** facilitates the development and maintenance of secondary sexual characteristics such as a deep voice, body and facial hair, and development of male reproductive structures. The testes also produce the hormone, **inhibin**, that inhibits the production of follicle stimulating hormone.

In females, luteinizing hormone stimulates the ovarian secretion of estrogens and the secretion of progesterone by the **corpus luteum**. **Follicle stimulating hormone** also stimulates the secretion of estrogens by the ovaries. These steroidal hormones, **estrogens** and **progesterone**, are responsible for the maintenance of female secondary characteristics and for the regulation of the menstrual cycle, pregnancy, and lactation. (Hormonal regulation of female reproduction is discussed with the reproductive system.) Inhibin, which inhibits the production of follicle-stimulating hormone, is also secreted by the ovaries.

Other organs have an endocrine function in addition to their primary function. The **thymus**, located on the midline in the anterior of the thoracic cavity, produces thymosin. **Thymosin** promotes the development of white blood cells, or lymphocytes, of the immune system. Specialized cells in the heart's atrial walls produce atrial natriuretic hormone, or atriopeptin. **Atriopeptin** decreases blood volume and pressure by increasing the excretion of sodium and, therefore, of water in urine. The stomach produces **gastrin** which stimulates production of hydrochloric acid and the digestive enzyme, pepsin. Mucosal cells of the small intestine secrete the hormones, gastric inhibitory peptides, secretin and cholecystokinin. Gastric inhibitory peptide inhibits the secretion of gastric juice and reduces the rate at which the stomach empties while stimulating the release of insulin from the pancreas. **Secretin** stimulates the pancreas to produce a stomach acid neutralizer. **Cholecystokinin** stimulates the gall bladder to release bile and the pancreas to release digestive enzymes. The **placenta** synthesizes and secretes **human chorionic gonadotropin hormone**, estrogen, and progesterone.

Chapter 2

THE BLOOD

Blood is classified as liquid connective tissue with three general functions, transportation, protection, and regulation. In blood, oxygen is transported from the lungs to the cells and carbon dioxide from the cells to the lungs. Blood also transports hormones and nutrients to body cells and removes metabolic wastes. The protective role of blood entails clotting to prevent blood loss and immune functions involving phagocytic white blood cells and plasma proteins. The regulatory roles of blood involve temperature and pH homeostasis. pH reflects the acidity of a fluid and has an impact on protein structure and enzyme activity. By circulating through skeletal muscles, blood removes heat which is then transported to other tissues and the skin where it is dissipated.

Blood is a viscous fluid composed of plasma and formed elements. It is purplish in color until exposed to air whereupon the color becomes red. Whole blood comprises 8% of body weight. The straw-colored plasma accounts for 55% of blood volume in males and 60 to 65% of blood volume in females; the remaining 35 to 45% being formed elements. **Plasma** is a dynamic solution of over 100 different organic and inorganic solutes dissolved in water. Water accounts for 91.5% of the weight of plasma while proteins account for 7% and solutes, 1.5%. The three main types of plasma proteins are albumins, 54%, globulins, including antibodies, 38%, and coagulation factors, such as fibrinogen, 7%.

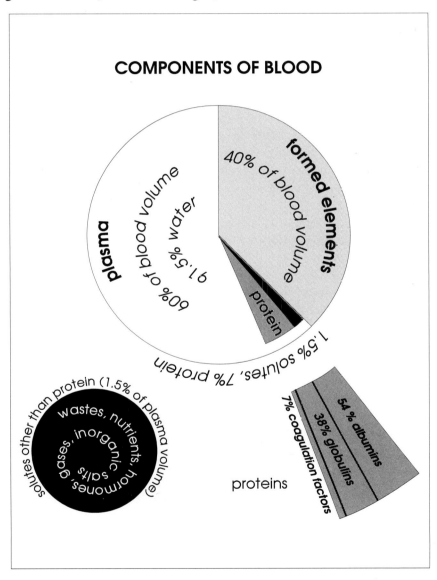

COMPONENTS OF BLOOD

Albumins are produced in the liver by hepatocytes. Albumins function in the maintenance of plasma osmotic pressure, blood volume, and the transport of lipids and other substances. **Alpha** and **beta globulins** transport lipids and fat-soluble vitamins which otherwise would not be soluble in water. One beta globulin, **fibrinogen**, is the precursor of fibrin which provides the framework for clot formation. **Gamma globulins** are immunoglobulins, the circulating antibodies that attack foreign proteins and pathogens. **Serum** is blood plasma with fibrinogen and other clotting factors removed.

The other solutes of plasma include electrolytes, respiratory gases, metabolic wastes, nutrients, and water-soluble vitamins and hormones. The electrolytes, or ions, of plasma are derived from the dissociation of inorganic and organic salts. Principle positive ions (cations), in order of abundance, are sodium, calcium, magnesium, and potassium. The principle negative ions (anions) are chloride, bicarbonate, protein anions, phosphate, organic acids, and sulfate. Uric acid and urea along with bilirubin and creatine are metabolic wastes that are transported to the kidneys for excretion. Nutrients in plasma are amino acids from protein digestion, glucose from carbohydrate digestion, and fatty acids and glycerol from lipid digestion. The gases, carbon dioxide, nitrogen, and oxygen are also present.

Carbon dioxide is a byproduct of catabolism that's released from the lungs. Only 7-10% of the carbon dioxide in plasma occurs as CO_2, approximately 70% occurs as bicarbonate ion, and the remaining 23% is carried in red blood cells, bound by the protein hemoglobin. The enzyme, carbonic anhydrase, catalyzes the formation of carbonic acid from water and CO_2. The acid can then dissociate to form bicarbonate and free hydrogen ions. Oxygen must be delivered to cells where it is required as the ultimate electron receptor in the energy-releasing reactions of electron transport. Only 3% of transported oxygen is dissolved in the plasma while 97% is bound with hemoglobin of red blood cells forming oxyhemoglobin. Plasma nitrogen has no known function.

The formed elements of blood are cells and their fragments which are suspended in the plasma. **Erythrocytes**, or red blood cells, are the most numerous, accounting for over 99% of the total and ranging in concentration from 4.5 - 7.0 million per milliliter. Erythrocytes are biconcave discs, 7 microns in diameter, thicker on the sides than in the middle. Their size, shape, and flexibility permit them to squeeze through capillaries. Erythrocytes lack a nucleus, ribosomes, mitochondria, and certain other organelles. Their metabolic activity is streamlined for oxygen and carbon dioxide transport between the lungs and body tissues. They are packed with hemoglobin molecules which account for 1/3 of the cell weight.

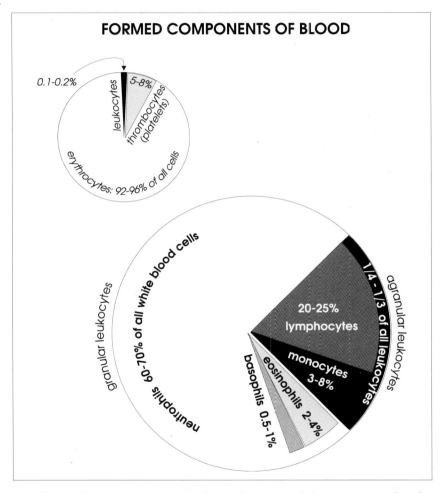

FORMED COMPONENTS OF BLOOD

Hemoglobin is composed of four iron-containing, non-protein, ring-like pigments called heme plus the protein, globin. Globin is actually four polypeptides, two alpha and two beta chains. An iron atom, Fe^{2+}, within each heme binds one oxygen molecule, forming oxyhemoglobin. As erythrocytes travel through the body, oxygen is released from the hemoglobin, diffusing into the interstitial fluid and then into cells. Once the oxygen is released, hemoglobin can transport CO_2 in the form of carbaminohemoglobin. Carbon dioxide is transported to the lungs where it is released then exhaled.

Red blood cells arise from stem cells, or hemocytoblasts, that develop into proerythroblasts. Proerythroblasts undergo a series of mitotic divisions before differentiating into reticulocytes, or immature red blood cells, which produce hemoglobin for incorporation into the plasma membrane. Maturation of reticulocytes begins with the expulsion of the nucleus afterwhich they enter circulation and mature, after one day, into erythrocytes. Hematopoiesis, or **hemopoiesis**, the formation of red blood cells, occurs in red bone marrow while **erythropoiesis**, the differentiation and maturation of erythrocytes, occurs in red bone marrow and blood.

The erythropoiesis stimulating hormone, **erythropoietin**, initiates the differentiation of stem cells into proerythrocytes and, finally, into reticulocytes. Erythropoietin comes primarily from the kidneys but also from the liver. The release of erythropoietin from kidney cells is triggered by hypoxia, an inadequate supply of oxygen that may derive from an insufficient number of red blood cells. Vitamin B_{12} and folic acid are necessary for erythrocyte production.

The average life span of red blood cells is 100 to 120 days. New cells are produced at a rate of more than two million per second to replace worn out cells that are destroyed by fixed phagocytic macrophages in the liver and spleen. The heme pigment of phagocytized erythrocytes is broken down into **bilirubin**, a bile pigment that is excreted. Iron from phagocytized red blood cells is transported by a plasma protein, **transferrin**, back to bone marrow for incorporation into new hemoglobin. Excess iron is stored as ferritin and hemosiderin in the liver, spleen, and muscles. The globin molecule of phagocytized red blood cells is catabolized to amino acids which are recycled for protein synthesis.

When whole blood is centrifuged in the presence of anticoagulants, like **heparin**, it is separated into layers with the most dense components on the bottom. The percentage of the total blood volume occupied by the bottom layer of red blood cells is called the **hematocrit**. The hematocrit averages 46% in men and 42% in women. The difference between sexes results from the positive effect of androgens and the negative effect of estrogens on red blood cell production. The rate of erythropoiesis or of hemopoiesis is measured by determining the percentage of reticulocytes in the blood, normally about 1%.

Blood type refers to the presence or absence of specific surface antigens in the erythrocyte cell membrane that are genetically determined glycoproteins and glycolipids. There are over 100 kinds of these antigens, called agglutinogens, that function in immunodefense. There are over two dozen blood type groupings based on these antigens. But three agglutinogens are of particular importance in blood typing, agglutinogens A, B, and Rh. Agglutinogens A and B characterize the ABO blood group system. The red blood cells of blood type A have A agglutinogen, cells of type B have B agglutinogen, cells of type AB have both A and B agglutinogens, while cells of type O have neither A nor B agglutinogen.

Three alleles of the single "I" gene determine **ABO blood type**. Allele I^A encodes the A agglutinogen, allele I^B encodes the B agglutinogen, and allele I^O encodes neither. Because we inherit two I-gene alleles, one from each parent, blood types are determined as follows: blood type AB with alleles I^A I^B, blood type A with allele I^A in combination with either I^A or I^O, blood type B with allele I^B in combination with either I^B or I^O, and blood type O with two I^O alleles.

Plasma contains naturally occurring agglutinins that can react with either A or B agglutinogens on the red blood cells. A person with type A agglutinogens will develop B agglutinins, or anti-B antibodies, which react with agglutinogen B. The converse is true for type B individuals who produce anti-A antibodies. A blood transfusion between blood type A and type B individuals would result in the clumping or agglutination of red blood cells. The resultant hemolysis, or rupturing of red blood cells, would damage the kidneys and may cause death. People with blood type AB are called universal recipients because they do not have either anti-A or anti-B antibodies and can, therefore, receive blood from all four blood types. Individuals with type O blood have no A or B agglutinogens with which antibodies would react and can, therefore, donate blood to individuals of any of the four blood types. Type O individuals are universal donors.

The **Rh blood group** is named for an antigen discovered in the Rhesus monkey. Individuals with Rh antigens are Rh^+ while those without the antigen are Rh^-. 85 to 88% of the population is Rh^+. Normally, neither Rh^+ or Rh^- individuals have anti-Rh agglutinins. However, after a transfusion of Rh^+ blood into an Rh^- individual, the immune system responds by making anti-Rh agglutinins. Similarly, development of hemolytic disease of the newborn, or erythroblastosis fetalis, results from the mixing of fetal Rh^+ blood with the mother's Rh^- blood during a previous pregnancy. Anti-Rh agglutinins developed by the mother pass through the placenta into the fetal blood where they react with the Rh agglutinogens resulting in agglutination and hemolysis of fetal red blood cells.

BLOOD TYPE PROBLEM

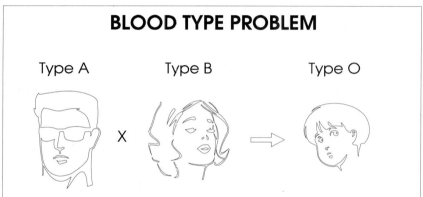

Type A   Type B   Type O

A man with type A blood and a woman with type B blood have a child with type O blood. What proportion of their children are expected to have a blood type matching one of their types?

The blood type O child has the genotype $I^O I^O$. Therefore both parents possess an I^O allele. The genotype of the father must be $I^A I^O$ while that of the mother must be $I^B I^O$.

Using a Punnet square reveals that 1/4 of the children are expected to have blood type A and 1/4 blood type B.

Therefore, 1/2 of the children are expected to have a blood type matching one of their parents.

mother's alleles

	I^B	I^O
I^A	$I^A I^B$	$I^A I^O$
I^O	$I^B I^O$	$I^O I^O$

father's alleles

When blood is centrifuged a thin white layer, the buffy coat, lies above the erythrocyte layer. It consists of **leukocytes**, or white blood cells, along with thrombocytes, or platelets. These cells account for less than 1% of the total blood volume. Leukocytes are nucleated and lack hemoglobin. Hemocytoblast stem cells may differentiate into myeloblasts which then mature into neutrophils, eosinophils, and basophils. Or, stem cells may give rise to either monoblasts from which monocytes arise or lymphoblasts from which lymphocytes arise. **White blood cells** are generally larger than red blood cells, varying in diameter from 10 to 20 microns. Leukocytes defend the body against bacteria, viruses, parasitic worms, toxins, and tumors.

Typically, there are 5,000 to 10,000 white blood cells per milliliter of blood. The two main groups of white blood cells are derived from hemocytoblast stem cells. **Granulocytes** have granules in their cytoplasm while agranulocytes do not. Among granulocytes there are neutrophils, eosinophils, and basophils while among **agranulocytes** there are monocytes and lymphocytes.

Neutrophils phagocytize pathogens and account for 60 to 70% of all leukocytes. Neutrophils are the first leukocytes to arrive at sites of tissue damage where they engulf bacteria. Their numbers increase dramatically during acute infections. Lymphocytes account for 20 to 25% of the white blood cell population. **Lymphocytes** produce antibodies and antimicrobial chemicals. **Monocytes**, representing 3 to 8% of all white blood cells, phagocytize pathogens in the blood. But, upon entering tissues, monocytes differentiate into the larger phagocytic macrophages. Monocytes also play a role in the interaction of leukocytes with antigens.

Only 2 to 4% of the leukocytes are eosinophils. **Eosinophils** participate in allergic responses and defend against parasitic worms. **Basophils** are the least numerous leukocytes, less than 1% of the total. Basophils secrete histamine and serotonin in allergic reactions, intensifying the inflammatory response.

Platelets, or thrombocytes, are fragments of much larger megakaryoblasts. In red bone marrow, a single megakaryoblast may give rise to thousands of platelets which then live for only 5 to 9 days. Platelets are disc-shaped and small, 2 to 4 microns in diameter. The concentration of platelets is 250,000 to 400,000 per milliliter in whole blood. They play a crucial role in blood clotting.

Blood escapes when a blood vessel is damaged. This hemorrhage will cause death if not stopped. **Hemostasis**, the prevention of blood loss, occurs in three stages. First, smooth muscle in the vessel walls spasm, causing vascular constriction that restricts blood flow. Vascular spasm lasts only a few minutes but allows platelets to respond. Second, platelets adhere to collagen fibers in the connective tissue and to each other, creating a platelet plug. Platelets also release serotonin and other chemicals that stimulate smooth muscle contraction and prolong vascular spasm. Third, blood clotting, or coagulation, occurs as clotting factors transform the blood from a liquid to a gel. Clotting factors are calcium ions and various proteinaceous procoagulants.

At the site of damage, the inactive protein clotting factor, **prothrombin**, is converted into thrombin. **Thrombin** converts the plasma protein, **fibrinogen**, into insoluble fibers of fibrin. The original platelet plug becomes surrounded and reinforced with a shield of fibrin fibers that entangle red blood cells, strengthening the patch in the vessel wall.

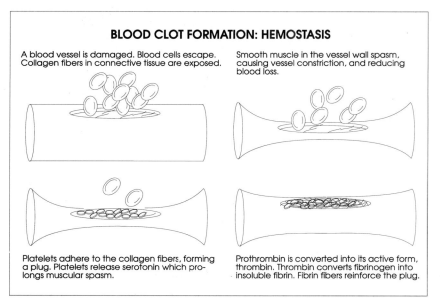

BLOOD CLOT FORMATION: HEMOSTASIS

A blood vessel is damaged. Blood cells escape. Collagen fibers in connective tissue are exposed.

Smooth muscle in the vessel wall spasm, causing vessel constriction, and reducing blood loss.

Platelets adhere to the collagen fibers, forming a plug. Platelets release serotonin which prolongs muscular spasm.

Prothrombin is converted into its active form, thrombin. Thrombin converts fibrinogen into insoluble fibrin. Fibrin fibers reinforce the plug.

Dissolution of the clot, or **fibrinolysis**, is a consequence of the conversion of plasminogen, a proenzyme, into plasmin. **Plasmin** digests fibrin strands, eroding the clot.

Most of the clotting factors, including prothrombin, require vitamin K for their synthesis in the liver. Thus, a vitamin K deficiency may cause clotting problems. **Hemophilia** is an inherited disorder characterized by inadequate production of clotting factors. Hemophilia may be controlled by transfusion of clotting factors.

Abnormal or excessive clotting can produce a drifting blood clot, or **embolus,** that can occlude downstream circulation. Loss of circulation leads to an infarct or tissue death. An infarct in the brain causes a stroke. In heart muscle, a myocardial infarction may cause death.

Chapter 3

THE HEART

Blood flows through pulmonary and systemic circuits that begin and end at the heart. The **heart** is a muscular pump located behind the sternum and on the diaphragm in a mid-anterior thoracic partition, the mediastinum.

The heart and a portion of the vessels attached to its base are enclosed by a double layered membrane, the pericardial sac or **pericardium**. The outer, fibrous pericardium is a tough membrane of fibrous connective tissue that protects the heart from overstretching during vigorous exercise while also anchoring the heart to the diaphragm and sternum. The inner, serous pericardium is composed of outer, parietal and inner, visceral layers. The parietal layer is fused to the fibrous pericardium while the visceral layer adheres tightly to the muscle of the heart. The visceral layer, or **epicardium**, is part of the heart wall. The small pericardial cavity lies between the parietal and visceral layers of the serous pericardium. The pericardial cavity contains a film of friction-reducing serous fluid.

The heart wall has three layers: the epicardium, myocardium, and endocardium. The middle layer, the **myocardium**, forms the bulk of the heart's cardiac muscle tissue, connective tissue, blood vessels, and nerves. Cardiac muscle tissue is arranged in layers of cardiac muscle cells which are small, cylindrical branching striated cells with one or two nuclei. Cardiac muscle cells are connected by intercalated disks which, with the branching structure of cells, allows cardiac muscle fibers to form networks. Consequently, when one fiber is electrically excited, the impulse of action potential is transmitted to neighboring fibers in all directions throughout the network.

The outer **epicardium** is a thin, transparent membrane surrounding the heart and attached to the myocardium. The innermost layer, the **endocardium** consists of a layer of simple squamous epithelium over a connective tissue layer, providing a smooth inner lining that facilitates the passage of blood. The endocardium is continuous with the lining of the blood vessels attached to the heart and forms the valves of the heart.

The internal cavity of the heart is organized into four chambers: the right and left atria, and the right and left ventricles. The **atria** are thin-walled chambers that receive blood returning to the heart from the veins. The two **ventricles** are thick-walled muscular pumps that squeeze blood out of the heart and into arteries.

The **right atrium** receives deoxygenated blood from organs through three veins, the superior vena cava, the inferior vena cava, and the coronary sinus. The **superior vena cava** receives blood from the head, neck, and upper extremities while the **inferior vena cava** receives blood from the lower extremities, and the pelvic, abdominal, and thoracic regions. The **coronary sinus**, on the posterior of the heart, collects blood draining from the myocardium. The **left atrium** receives oxygenated blood directly from the lungs via two right and two left pulmonary veins.

Each atrium has a small, wrinkled **auricle** extending anteriorly that permits the receipt of a greater volume of blood. A connective tissue layer, the **interatrial septum**, separates the two atria. An oval depression within the interatrial septum, the **fossa ovalis**, is the remnant of the foramen ovale, an opening in the interatrial septum of the fetal heart. The **foramen ovale** usually closes soon after birth. Externally, a groove, the **coronary sulcus**, separates the atria from the ventricles.

The **right ventricle** pumps deoxygenated blood to the lungs through the pulmonary trunk. The **left ventricle** pumps oxygenated blood into the aorta and, ultimately, to all organs of the body. The internal walls of the ventricles have ridges and folds, the **trabeculae carnae**, and projecting, cone-like papillary muscles. Internally, the ventricles are separated by a thick interventricular septum. Externally, anterior and posterior interventricular sulci separate right and left ventricles. The sulci contain coronary blood vessels and fat deposits.

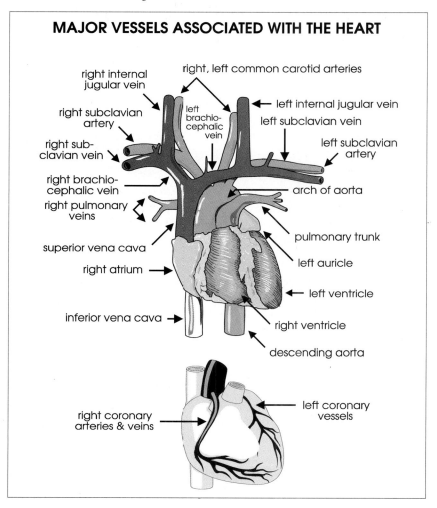

MAJOR VESSELS ASSOCIATED WITH THE HEART

There are four one-way valves in the heart. Two of these, the bicuspid and tricuspid valves are atrioventricular valves that connect atria with their adjacent ventricles. **Atrioventricular valves** consist of folds of cusp-forming endocardium and a fibrous connective tissue ring attached to the atria and ventricles. The **bicuspid**, or left atrioventricular, valve has two cusps that separate the left atrium and left ventricle. The **tricuspid**, or right atrioventricular, valve has three cusps separating the right atrium and right ventricle.

The left and right atrioventricular valves open and close in response to pressure changes on either side. Valves are attached by minute tendons, the chordae tendineae, to papillary muscles that regulate the tension in the tendons. Atrioventricular valves prevent back flow because as the ventricle contracts, the papillary muscles simultaneously contract, tightening the chordae tendineae and preventing the valve cusps from turning inside out.

The other one-way valves are the two **semilunar valves**. One, the **aortic valve**, separates the left ventricle from the aorta. The other, the **pulmonary valve**, separates the right ventricle from the pulmonary trunk. The semilunar valves are composed of three cup-like cusps of fibrous connective tissue and endothelium and prevent back flow into the ventricles. After contraction, the ventricles relax, causing the pressure in the ventricular spaces to drop and causing blood flow back toward the heart. This blood enters the cups of the valve cusps, causing them to fill and bulge toward the center of the artery, thereby occluding it.

The right and left coronary arteries supply blood to the heart. The **coronary arteries** branch off of the aorta and encircle the heart, lying in the atrioventricular groove. The left coronary artery, which divides into the anterior interventricular artery and the circumflex artery, supplies the left side of the heart. The anterior interventricular artery follows the anterior interventricular sulcus to supply the interventricular septum and anterior walls of both ventricles. The circumflex artery supplies the left atrium and posterior wall of the left ventricle.

The right coronary artery which branches into the marginal and posterior interventricular arteries supplies the right side of the heart. The marginal artery supplies the lateral side while the posterior interventricular artery supplies the posterior ventricular walls. Near the apex of the heart the anterior and posterior interventricular arteries anastomose or merge.

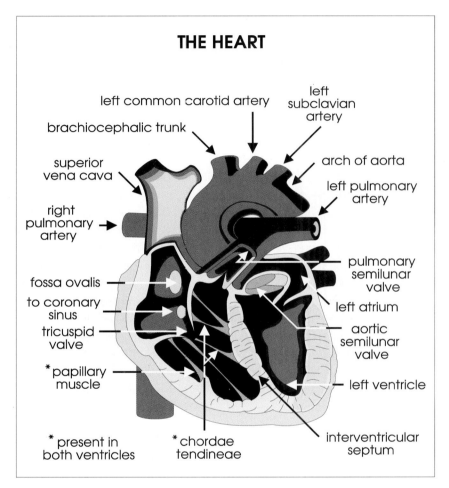

THE HEART

left common carotid artery

left subclavian artery

brachiocephalic trunk

superior vena cava

arch of aorta

left pulmonary artery

right pulmonary artery

fossa ovalis

to coronary sinus

tricuspid valve

*papillary muscle

pulmonary semilunar valve

left atrium

aortic semilunar valve

left ventricle

* present in both ventricles

*chordae tendineae

interventricular septum

After blood passes through myocardial capillaries it enters the cardiac or coronary veins which follow roughly the same paths as the coronary arteries before flowing into the coronary sinus that empties into the right atrium. The **coronary sinus** receives blood from: (1) the great cardiac vein that runs along the anterior interventricular sulcus, draining the anterior region of the heart, (2) the middle cardiac vein that runs in the posterior interventricular sulcus, draining the posterior region of the heart, and (3) the small cardiac vein that courses along the heart's right inferior margin, draining the right inferior region.

   Myocardial ischemia is reduced blood flow to the heart that causes
hypoxia, a reduction in oxygen supply that weakens muscle cells. **Angina
pectoris** is severe pain that accompanies myocardial ischemia. If a branch
of the coronary artery becomes blocked such that blood supply is cut off,
cardiac muscle cells die from lack of oxygen, resulting in a **myocardial
infarction**, or heart attack.

   Heart contractions are synchronized so that atria contract simultaneously, followed by the ventricles. Synchronized contractions are coordinated by specialized cardiac muscle cells, the auto-rhythmic or
self-excitable cells. The network of auto-rhythmic cells starts each heart
contraction and provides for rapid, coordinated spread of excitation. The
basic rhythm of the heartbeat is established by the pacemaker, or **sinoatrial
node**, a small mass of cells embedded in the right atrial wall near the opening of the superior vena cava. The sinoatrial node initiates impulses at a rate
of approximately 75 per minute. Because heart muscle cells are connected
by cytoplasmic strands or gap junctions, electrical impulses easily spread
to adjacent cells. Thus, impulses from the sinoatrial node rapidly travel
throughout the atrial myocardium, causing the atria to contract simultaneously. Consequently, blood is pumped through the bicuspid and tricuspid
valves and into the ventricles. Some blood flow to the ventricles is the result
of passive filling. Here, blood flows from the atria due to the vacuum within
empty ventricles.

   Once the impulse reaches the border between atria and ventricles, its
continued conduction is prevented by a band on nonconducting fibrous tissue. To pass the fibrous block the impulse must travel through specialized
fibers of the atrioventricular node in the interatrial septum near the tricuspid
valve. This routing of the electrical transmission takes about one-tenth of a
second, giving the atria time to complete their contractions before ventricular contractions begin.

   The atrioventricular node connects with the large fibers of the
atrioventricular bundle, or **bundle of His**, which branches into left and right
bundles. These bundles conduct impulses very rapidly along the interventricular septum to the apex of the ventricles. Each bundle branches into
smaller conduction myofibers, or **Purkinje fibers**, that ramify throughout
the ventricular muscle. Thus, with each heartbeat a wave of depolarization
spreads through the atria to the atrioventricular node, then down the interventricular septum, and, finally, back from the heart's apex to its base.

An **electrocardiogram**, the ECG or EKG, records the electrical activity of the heart which can be monitored on the surface of the body. A normal EKG consists of deflections, called waves, that correlate with muscle depolarization and repolarization. The **P wave** is a small upward deflection produced by the depolarization of the atria just prior to their contraction. The large **QRS wave** complex deflection represents ventricular depolarization. Ventricular contraction begins after the **R wave** peak. A smaller **T wave** then corresponds to ventricular repolarization. The time between waves is also recorded. EKGs are useful in detecting and diagnosing cardiac arrhythmias which are abnormal patterns of contraction.

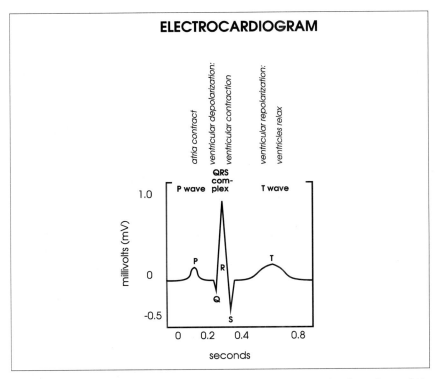

A **cardiac cycle** is the alternating contraction and relaxation of the atria and ventricles. **Systole** is the period of atrial or ventricular contraction when blood is expelled while **diastole** is the period of relaxation when the chambers again fill with blood. The cardiac cycle begins with atrial systole, when both ventricles are in diastole. Atrial contraction forces the atrioventricular valves open, allowing blood to flow into the diastolic ventricles. At rest, atrial systole lasts about one-tenth of a second while atrial diastole lasts for seven-tenths of a second. Ventricular systole lasts for three-tenths of a second. As pressure in the ventricles increases, the atrioventricular valves close and semilunar valves open allowing blood flow into the aorta and pulmonary artery. During ventricular systole the atria are in diastole, filling with blood returning via the superior and inferior vena cavae and the coronary sinus.

Following ventricular systole there is a four-tenths second period during which the atria and ventricles are in simultaneous diastole with the semilunar valves closed and the atrioventricular valves open. At this time, blood enters both the atria and ventricles but the ventricles are not fully filled until atrial systole.

Aspects of the cardiac cycle can be monitored with a stethoscope which is used to listen to heart sounds or **auscultations**. A low-pitched "lubb" is heard as the atrioventricular valves close when the ventricles begin to contract. A higher-pitched "dupp" is heard as the semilunar valves close. The time between the lubb and dupp is the period of ventricular systole. Because diastole lasts longer than systole, there is a pause between each lubb-dupp. The stethoscope aids in diagnosing mitral valve prolapse, a condition in which the bicuspid valve cusps do not close properly. **Heart murmurs** are abnormal sounds that may indicate problems with valve closure.

Cardiac output is the volume of blood in milliliters pumped by each ventricle per minute. Cardiac output is the product of heart rate in beats per minute and stroke volume in milliliters. The average heart rate is 75 beats per minute and the average stroke volume is 70 milliliters, giving an average cardiac output of 5.25 liters per minute. Because the total volume of blood is about 5 liters, the entire volume of blood is pumped through systemic and pulmonary circulations about once a minute.

Stroke volume depends on the amount of blood collected in the ventricles during diastole, called the end-diastolic volume, minus the volume remaining after contraction, called end-systolic volume. Factors that regulate stroke volume include: (1) preload, the degree of stretch of the cardiac muscle fibers just before contraction, (2) contractility, the force of contraction, and (3) afterload, the back pressure exerted by arterial blood. Preload is proportional to the force of contraction, a relationship known as the Starling law of the heart. Contractility is increased by the neurotransmitter, norepinephrine, which is released from postganglionic sympathetic neurons.

Heart rate is regulated by the autonomic nervous system and the hormones, epinephrine and norepinephrine, which are released by the adrenal medulla. Neural control originates in the cardiac center of the medulla oblongata and entails both sympathetic and parasympathetic branches. Specialized sensory receptors such as chemoreceptors that monitor blood chemistry, proprioceptors that monitor the positions of limbs and muscles, and baroreceptors that monitor blood pressure provide input into the cardiac center.

Stimulation by sympathetic nerves increases heart rate while stimulation by parasympathetic nerves decreases heart rate. Factors that influence heart rate include physical or emotional stress, hypercalcemia which is elevated plasma calcium, alkalosis which is elevated pH, acidosis which is depressed pH, hypoxia which is low plasma oxygen, elevated blood CO_2 concentration, and elevated body temperature.

Chapter 4

THE CIRCULATORY SYSTEM

The blood flows through the pulmonary and systemic circuits which are one-way circulatory routes that begin and end at the heart. The **pulmonary circulation** begins in the right ventricle which pumps deoxygenated blood to the lungs for oxygen absorption and carbon dioxide release. Oxygenated blood is carried back to the left atrium of the heart. The **systemic circulation** begins with the left ventricle which pumps the oxygenated blood to the tissues and returns deoxygenated blood to the right atrium of the heart. Systemic circulation delivers oxygen and blood to cells and transports metabolic wastes to the kidneys for excretion. Also, carbon dioxide is carried to the lungs for expiration.

Systemic arteries carry oxygenated blood to body tissues while **pulmonary arteries** carry deoxygenated blood to the lungs. **Systemic circuit veins** return deoxygenated blood to the heart while **pulmonary veins** carry oxygenated blood from the lungs to the heart.

Oxygenated blood leaves the heart in large elastic systemic arteries that then branch repeatedly into smaller and smaller arteries, then into smaller arterioles, and finally into capillaries. **Capillaries**, which are the smallest and most numerous of the blood vessels, form a connection between arteries and veins. **Venules** are the smallest veins from which blood flows into progressively larger veins until reaching the heart.

The tubular blood vessels, arteries, capillaries, or veins, have walls of three tissue layers: an inner tunica interna, a middle tunica media, and an outer tunica externa. The central cavity of vessels is the **lumen**.

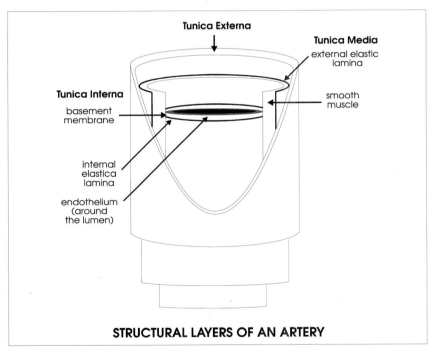

STRUCTURAL LAYERS OF AN ARTERY

The outer **tunica externa**, or tunica adventitia, is a layer of connective tissue with collagen and elastin fibers. An external elastic lamina separates the outer tunica externa from the middle tunica media in medium sized muscular arteries. Connective tissue fibers of the tunica externa usually anchor blood vessels to surrounding tissues. The tunica externa in large arteries and veins is itself supplied with oxygen and nutrients via small capillaries, the **vasa vasorum**.

The **tunica media,** usually the thickest layer, is primarily concentric layers of smooth muscle cells with interspersed elastin fibers. Loose connective tissue with collagen and elastic fibers binds the tunica media to both the outer tunica externa and the inner tunica interna. Smooth muscle fibers of the tunica media can contract, decreasing the vessel's diameter, restricting blood flow, and increasing blood pressure.

The inner **tunica interna** consists of an endothelium of precisely fitted simple, squamous epithelial cells. A basement membrane separates the tunica interna from the tunica media. An internal elastic lamina is also present in arteries.

Arteries are elastic or muscular. Elastic or conducting arteries, like the aortic and pulmonary trunks, the common carotid and subclavian arteries, are large vessels with diameters up to 2.5 cm. The highly resilient walls of these large arteries have a high density of elastic fibers but relatively few smooth muscle fibers in the tunica media. Muscular, or distribution, arteries with diameters of about four-tenths of a centimeter distribute blood to organs and skeletal muscles. The tunica media of muscular arteries has a large amount of smooth muscle fibers.

Arterioles, capillaries and venules lack the well developed tunica externa and media characteristic of larger blood vessels. **Arterioles** carry blood from arteries to capillaries. Their walls become more muscular and less elastic at their capillary ends which consist of endothelium cells, basement membrane and some smooth muscle. Arterioles regulate arterial pressure and control blood distribution to organs. A single arteriole usually gives rise to several capillaries which empty into venules.

Capillaries connect arterioles to venules. They are the only blood vessels whose walls permit gas, nutrient, and waste exchange between blood and surrounding interstitial fluids. A typical capillary is an eight-micron wide endothelial tube within a basement membrane. Capillaries occur in interconnected capillary beds. Skeletal muscles, the liver, kidneys and other large organs all have extensive capillary networks. Blood flow from arterioles into capillaries is controlled by smooth muscles or precapillary sphincters. When precapillary sphincters contract, metarterioles shunt blood across the capillary bed directly to the venules.

Veins have a thinner tunica interna and media than arteries, but a thicker tunica externa. Because blood pressure is low in veins, one-way valves are present in medium-to-large veins to prevent blood back flow.

Each bowl-shaped valve is a fold of the tunica media and interna with its concave surface pointing toward the heart. Venous sinuses are thin endothelial vessels without smooth muscle that are located in the coronary sinus of the heart and the intracranial vascular sinuses that drain blood from the brain.

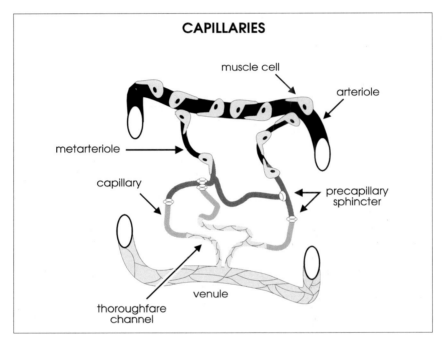

Simple diffusion is the basis for most capillary exchange. Oxygen, glucose, amino acids, hormones, and other metabolites diffuse down concentration gradients through capillary walls and into the interstitial fluid that surrounds tissue cells. Carbon dioxide and other metabolic wastes diffuse from tissue cells into the interstitial fluid and from there into capillaries. Vesicular transport of large molecules entails endocytosis and exocytosis. Bulk flow involves filtration of plasma under pressure through the capillary wall near the arteriole and absorption by tissue cells of ions, molecules, and other dissolved substances. Bulk flow is from a region of higher to lower pressure.

The blood pressure in capillaries is called **blood hydrostatic pressure**. It forces fluid from the capillaries into interstitial spaces. Hydrostatic pressure is highest at the arteriole end of a capillary. Osmosis accounts for the reabsorption of water from interstitial fluids. Osmotic pressure is the force of water movement from an area of lower to higher solute concentration. The osmotic pressure of blood is called the **blood colloid osmotic pressure** and depends on the difference in solute concentration between blood plasma and interstitial fluid. Water moves from interstitial fluid to capillary lumen until the osmotic pressure is equally opposed by hydrostatic pressure or until solute concentrations are equalized between blood plasma and interstitial fluid. Fluid exchange supplies oxygen and nutrients to the interstitial fluid and tissue cells and pulls carbon dioxide and metabolic wastes out of the cells and into the capillaries.

Reabsorption occurs primarily at the venule end of a capillary because that is where blood hydrostatic pressure is lowest. Thus, capillary exchange entails a microcirculation driven by hydrostatic and osmotic pressures. However, about 15% of the fluid entering the capillaries is not reabsorbed. This fluid, called **lymph**, is collected by lymphatic capillaries which then transfer it to lymph vessels. Lymph then empties into large veins near the heart. **Edema** is the local accumulation of interstitial fluid resulting in swelling, following injury or disease.

When at rest, about 62% of the 5-liter blood volume is in systemic veins and venules, 15% is in the systemic arterial system, 8% is in the heart, 10% is in the pulmonary system, and 5% is in capillaries. Thus, veins and venules are called **blood reservoirs**. Venous reservoirs in the liver, bone marrow, and skin hold about one-third of the venous volume. In the systemic system, blood is distributed as follows: one and one-half liters in the gastrointestinal tract, one liter in the kidneys, one liter in skeletal muscles, three-fourths liter in the brain, one-fourth liter in the heart, and one-half liter in the skin. After eating, a greater proportion of the blood goes to the gastrointestinal tract while during physical exercise a greater proportion goes to skeletal muscles. But, the amount of blood flow to the brain is constant.

Blood pressure is the force blood exerts on the walls of blood vessels and usually refers to arterial pressure which averages about 93 millimeters of mercury (mm Hg). But, blood pressure is only roughly 35 mm Hg as it enters the arteriole end of the capillaries. Capillary pressure declines from 35 mm Hg at the arteriole end to about 16 mm Hg at the venule end. Venous pressure is low, ranging from about 16 mm Hg in the venules to almost 0 at the right ventricle.

As oxygenated blood flows from the heart it passes from arteries of larger cross-sectional area into branches that are of smaller cross-sectional area. The branching culminates in capillaries which have the smallest cross-sections of all blood vessels. But, the sum of the cross-sectional area of all vessels increases with distance from the heart. Consequently, and in accordance with principles of fluid dynamics, blood pressure and velocity of flow decrease in the passage from large arteries to capillaries.

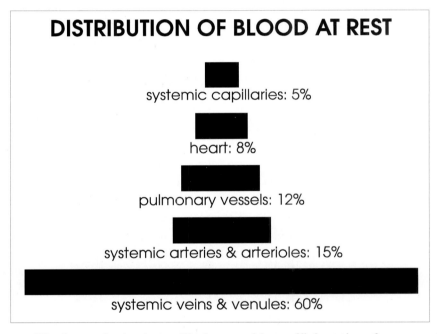

DISTRIBUTION OF BLOOD AT REST

systemic capillaries: 5%

heart: 8%

pulmonary vessels: 12%

systemic arteries & arterioles: 15%

systemic veins & venules: 60%

The low velocity in capillaries provides sufficient time for gas and nutrient exchange between the blood and interstitial fluid and tissue cells. The velocity gradually increases again as blood enters and moves through the venous system toward the heart.

Resistance to blood flow arises primarily from friction between blood and the walls of blood vessels. The amount of friction is proportional to blood viscosity and vessel length and inversely proportional to the vessel's cross-sectional area. Blood viscosity refers to the fluid thickness and depends largely on the number of red blood cells in a given volume of plasma.

Venous blood flow occurs when skeletal muscles contract, compressing the veins which run through them and forcing blood toward the heart. When muscles relax the milking action stops but the closing of venous valves prevents back flow. Venous blood flow also occurs during inspiration. When breathing in, movement of the diaphragm causes a decrease in thoracic cavity pressure and an increase in the abdominal cavity pressure. The resulting compression of abdominal veins pushes blood into thoracic veins. During expiration the relative pressures reverse but, again, valves prevent back flow. Exercise enhances blood flow to the heart due to both increased muscular contraction and a higher rate of breathing.

Blood pressure and flow are regulated by direct neural and hormonal controls. A group of neurons dispersed within the medulla oblongata of the brain stem, the vasomotor center, regulates blood vessel diameter. Sympathetic nerves of the autonomic nervous system release norepinephrine which diffuses to the smooth muscle cells in the tunica media causing vasoconstriction, muscular contraction that constricts the vessel lumen. Vasoconstriction increases resistance and blood pressure and decreases blood flow. Other neurons cause smooth muscles to relax, increasing the diameter of the lumen. This vasodilation decreases resistance and blood pressure and increases blood flow. Sensory input to the vasomotor center originates in sensory nerve endings in the walls of arteries. Sensory receptors include chemoreceptors that monitor CO_2 levels and blood acidity, and baroreceptors that monitor blood pressure.

Hormonal regulation of blood pressure involves epinephrine and norepinephrine from the adrenal medullae. These hormones increase blood pressure by stimulating cardiac output and vasoconstriction of some arterioles. Antidiuretic hormone (ADH) is released by the posterior pituitary in response to low blood pressure. ADH causes vasoconstriction and retention of water by the kidneys. The kidneys release renin in response to low blood pressure. Release of renin ultimately causes the production of angiotensin II, a hormone that stimulates vasoconstriction. Erythropoietin is also released by the kidneys in response to low blood pressure and stimulates the production of red blood cells that, in turn, increases blood volume. Atrial natriuretic peptide is produced by the cardiac muscle cells located in the right atrium in response to the excessive stretching during diastole that accompanies high blood pressure. Atrial natriuretic peptide lowers blood pressure and blood volume by stimulating water loss in the kidneys and peripheral vasodilation.

Circulation is evaluated by pulse and blood pressure. **Pulse rate** is the same as the heart rate. A pulse results from the expansion and recoil of elastic arteries as a consequence of changes in blood pressure between diastole and systole. The pulsing is most easily felt at superficial arteries like the radial artery of the wrist or the temporal artery which is lateral to the eye. The resting pulse rate is typically 70 to 80 beats per minute. **Tachycardia** is when the heart is beating quickly, over 100 beats per minute, while **bradycardia** is when the heart is beating slowly, under 60 beats per minute.

Blood pressure is usually measured by a **sphygmomanometer** which includes a pressure cuff and a pressure meter. The cuff is placed over the left brachial artery at the medial side of the biceps brachii muscle. The cuff, wrapped around the arm, is inflated until the pressure it exerts exceeds the pressure in the artery, resulting in arterial compression that prevents blood flow. A stethoscope is used to detect the sound of blood passing through the artery once the pressure exerted by the cuff is released. The pressure at which sound occurs is systolic blood pressure, the highest pressure that results from ventricular contraction. As the cuff pressure continues to fall, the sound suddenly becomes faint. The pressure at this point is the diastolic blood pressure, the lowest blood pressure in the arteries which occurs during ventricular relaxation. **Normal blood pressure** is about 120 millimeters of mercury systolic and 80 millimeters of mercury diastolic, or 120 over 80. **Pulse pressure** is the difference between systolic and diastolic pressure.

There are three circulatory routes, pulmonary circulation, systemic circulation, and fetal circulation. **Pulmonary circulation** carries deoxygenated blood from the right side of the heart. Deoxygenated blood enters through the right ventricle through the tricuspid valve. Blood is pumped through the pulmonary semilunar valve into the pulmonary trunk. The pulmonary trunk branches into right and left pulmonary arteries that enter the right and left lungs, respectively. In the lungs, these arteries branch into smaller and smaller vessels. Lung capillaries form a network around the air-filled sacs, or alveoli, permitting exchange of oxygen and carbon dioxide by diffusion. Freshly oxygenated blood enters venules that feed into larger veins. Two pulmonary veins emerge, one from the right and one from the left lung, and carry blood to the left atrium.

The **systemic circulation** begins in the left side of the heart. Blood passes from the left atrium, through the bicuspid valve, into the left ventricle. As the ventricle contracts, blood passes through the aortic semilunar valve into the aortic arch. The ascending part of the aortic arch curves over the superior surface of the heart while the descending part passes behind the heart before branching.

Typically, three arteries branch off of the aortic arch: the **brachio-cephalic**, the left common carotid, and the subclavian. A right common carotid artery branches off of the brachiocephalic artery and with the left **common carotid** supplies the brain. The common carotids branch into the right and left **internal carotids** and the right and left **external carotids**. The external carotids supply the anterior portion of the face and neck, sides of the head, skull, scalp, and dura mater. The internal carotids supply the brain, middle ear and pituitary gland. Like the carotids and their branches, most blood vessels are right and left paired.

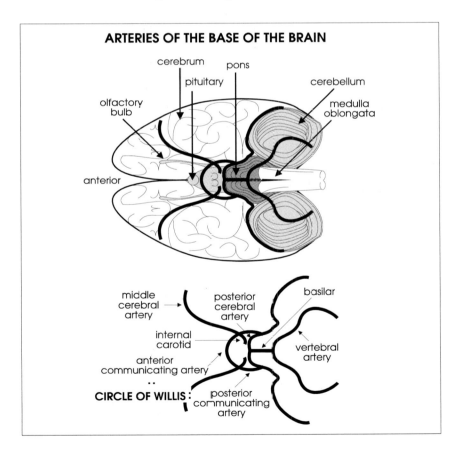

ARTERIES OF THE BASE OF THE BRAIN

The right and left **subclavian arteries** supply the neck, muscles of the upper arm and shoulder, thoracic wall, spinal cord, and brain. Coming off the subclavians are the right and left **axillary arteries** that, along with their branches, supply the pectoral muscles and muscles of the shoulder and upper arm. The **brachial arteries** are branches of the axillaries that supply muscles of the shoulder, arm, forearm, and hand. The brachial splits at the elbow, becoming the **ulnar** and **radial arteries**. The **vertebral arteries** also originate off of the subclavian arteries to supply the spinal cord, cerebellum, brain stem, and deep muscles of the neck. The **basilar** arises from a junction of the right and left vertebral arteries to supply the pons, cerebellum, and internal ear.

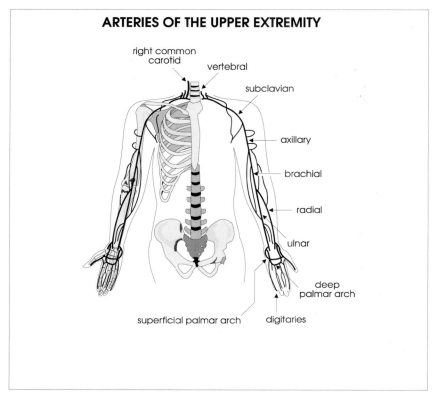

ARTERIES OF THE UPPER EXTREMITY

right common carotid
vertebral
subclavian
axillary
brachial
radial
ulnar
deep palmar arch
superficial palmar arch
digitaries

The **thoracic aorta** is a continuation of the aortic arch. As it descends, visceral branches supply various organs while parietal branches supply the body wall. The **abdominal aorta**, a continuation of the thoracic aorta, also sends off visceral and parietal branches.

The first major visceral branch is the **celiac trunk** which gives rise to the **common hepatic artery**, the splenic artery, and the left gastric artery. The common hepatic artery has three branches. One, the **hepatic artery** proper supplies the liver and gallbladder. Another, the **right gastric artery**, supplies the stomach and duodenum. The last branch, the **gastroduodenal artery**, supplies the stomach, duodenum, and pancreas.

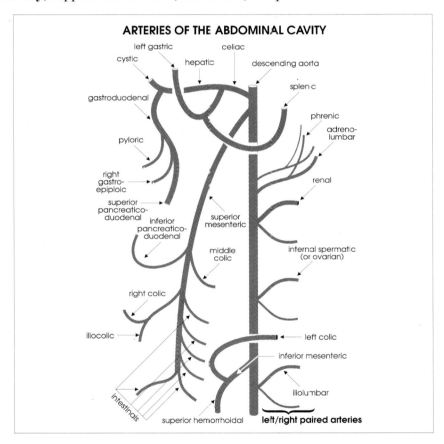

ARTERIES OF THE ABDOMINAL CAVITY

The **splenic artery** supplies the spleen and has three branches. These are the **pancreatic artery** that supplies the pancreas, and the short **gastric** and **left gastroepiploic arteries** that both supply the stomach. The other branch of the common hepatic, the **left gastric**, also supplies the stomach.

The second major branch of the abdominal aorta is the superior mesenteric artery. Branches of the **superior mesenteric** supply the small intestine, cecum, ascending and transverse colon, and pancreas. Next, the abdominal aorta gives off three paired branches. These are the **suprarenal arteries** that supply the adrenal glands, the **renals** that supply the kidneys, and the **gonadals** that supply the gonads.

The **inferior mesenteric** is a branch off of the aorta that supplies the last third of the large intestine including the transverse, descending, and sigmoid colon, and rectum. Other tributaries, the **lumbar arteries**, supply the vertebrae, spinal cord and its meninges, and the muscles and skin of the lumbar region of the back.

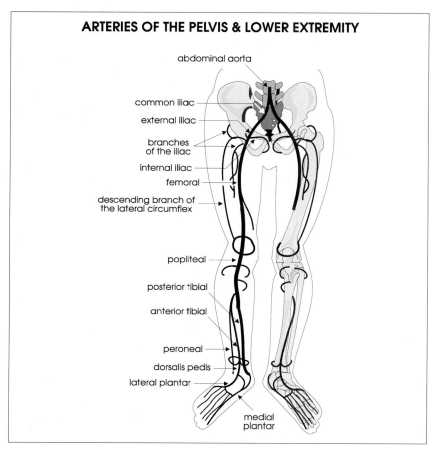

ARTERIES OF THE PELVIS & LOWER EXTREMITY

- abdominal aorta
- common iliac
- external iliac
- branches of the iliac
- internal iliac
- femoral
- descending branch of the lateral circumflex
- popliteal
- posterior tibial
- anterior tibial
- peroneal
- dorsalis pedis
- lateral plantar
- medial plantar

At its terminus, the abdominal aorta splits into the right and left **common iliac**. Branches of the iliac arteries supply the pelvis and lower extremities. Each common iliac divides to form a large **external iliac artery** and a smaller **internal iliac artery.** The internal iliac enters the pelvic cavity to supply the urinary bladder, the pelvis, the external genitalia, medial aspect of the thigh, and in females, the uterus and vagina. The external iliac arteries supply the lower limbs. In the thigh, the external iliac divides into **femoral** and **deep femoral arteries**. Both femoral arteries supply thigh muscles, the external genitals, the abdominal wall, and groin.

The femoral arteries run to the medial and posterior side of the thigh, then posterior to the knee as the **popliteal artery**. The popliteal supplies the muscles in the region of the knee and the bones, the femur, patella, and tibia. The popliteal divides into **posterior tibial** and **anterior tibial arteries** that supply muscles and bones of the lower leg. At the ankle, the anterior tibial artery continues as the **dorsalis pedis** that supplies the ankle and foot. The **peroneal artery** branches off the posterior tibial artery to supply the medial side of the fibula and the calcaneus of the foot. Medial and lateral **plantar arteries** also branch from the posterior tibial to feed the arch of the foot.

The major veins of the body are the **superior vena cava** which drains the head and upper extremities and the **inferior vena cava** which drains the torso and lower extremities. These veins empty into the right atrium. Superficial veins empty into deep veins. Where there are arteries, there are usually corresponding veins with the same name.

In the head, **intracranial vascular sinuses** are located between layers of the dura mater and receive blood from the brain. These are the sigmoid, superior and inferior sagittal, and transverse sinuses. The right and left **internal jugular veins** are a continuation of the sigmoid sinuses and drain the brain, face, and neck while the **external jugular veins** drain blood from the facial muscles, scalp, and parotid glands. The internal jugular veins descend on either side of the neck where they join with the right and left **subclavian veins** to form the right and left **brachiocephalic veins** which empty into the superior vena cava. External jugular veins run inferiorly in the neck and drain into the subclavian veins. The **vertebral veins** drain the posterior regions of the brain and descend through the transverse foramina of the cervical vertebrae and empty into the subclavian veins.

ARTERIAL DISTRIBUTION IN MAJOR BODY REGIONS

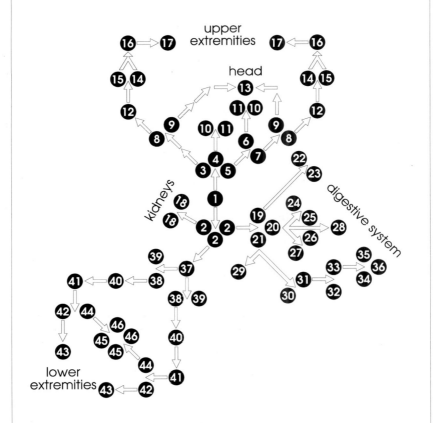

KEY TO ARTERIAL DISTRIBUTION

upper extremities

14- radial   15- ulnar   16- palmar arches   17- digitals

12- brachial   13- basilar

8- axillary   9- vertebral   10- external carotid   11- internal carotid

5- brachiocephalic   6- right common carotid   7- right subclavian

3- left subclavian   4- left common carotid

1- arch of the aorta

digestive system

2- descending aorta

25- inferior pancrecticoduodenal   26- right colic   27- iliocolic   28- intestinals

22- superior hemorrhoidal   23- left colic   24- middle colic

19- inferior mesenteric   20- superior mesenteric   21- celiac

34- pyloric   35- right gastroepiploic   36- superior pancreaticoduodenal

29- left gastric   30- splenic   31- hepatic   32- cystic   33- gastroduodenal

kidneys

18- renal

lower extremities

37- common iliac   38- external iliac   39- internal iliac

40- femoral   41- popliteal   42- anterior tibial   43- dorsalis pedis

44- posterior tibial   45- peroneal   46- plantars

Venous return from the upper extremities is from deep veins that usually accompany the arteries and from superficial veins under the skin. **Major superficial veins**, which are often visible, are the cephalic, basilic, median antebrachial, and median cubital vein. The **cephalic vein** ascends along the radial side of the forearm then goes deep into the shoulder region, emptying into the subclavian vein. The basilic and median antebrachial veins ascend on the ulnar side of the forearm. The **basilic vein** joins the brachial vein in the axillary region, forming the **axillary vein** which also empties into the subclavian vein. The **median antebrachial** drains the palmar venous arch of the hand and empties into the **median cubital vein** which is prominent on the anterior surface of the arm at the elbow. The prominence of the median cubital renders it the preferred site for venous punctures. The median cubital ascends from the cephalic vein to the basilic vein.

Blood from the lungs, esophagus, pericardium, vertebrae, diaphragm, and thoracic spinal cord is drained by the **azygos system**, a network of veins on each side of the vertebral column that includes the **azygos**, **hemiazygos**, and **accessory azygos veins**. The azygos system empties primarily into the superior vena cava but some branches empty into the inferior vena cava. Blood from the diaphragm, abdominal, and pelvic regions is returned via the inferior vena cava. Veins which drain the abdomen are the lumbar veins, gonadal veins, renal veins, suprarenal veins that are associated with the adrenal glands, phrenic veins of the diaphragm, and hepatic veins from the liver. The **inferior vena cava** receives blood from the lower limbs and liver by way of **hepatic veins**.

The **internal iliac vein** drains the gluteal muscles, urinary bladder, and rectum. And, in males it drains the prostate gland and vas deferens while in females it drains the uterus and vagina. The **external iliac veins** are a continuation of the **femoral veins** which receive blood from the lower limbs. The internal and external iliac veins join to form the **common iliac**, which is continuous with the inferior vena cava.

Like the upper extremities, **venous return in the lower extremities** involves a deep network that follows the arteries and a superficial network. The **great saphenous**, the longest in the body, is a superficial vein beginning at the medial end of the dorsal venous arch of the foot. The great saphenous ascends anteriorly to the medial malleolus of the fibula then along the medial aspect of the leg and thigh until it merges with the femoral vein in the groin. The great saphenous vein is used for grafts in coronary artery bypass surgery and as a site for long-term administration of intravenous fluids. The **small saphenous**, on the lateral and posterior side of the lower extremity, stops at the knee where it merges with the popliteal vein.

The **deep veins of the leg** are the anterior and posterior tibial veins, the popliteal vein, and the femoral vein. The **anterior tibial vein**, the superior continuation of the dorsalis pedis vein of the foot, drains the foot and anterior muscles in the leg. The **posterior tibial vein** receives blood from the medial and lateral **plantar veins** of the foot and from muscles of the posterior of the leg. The **peroneal vein** drains the lateral muscles of the leg, emptying into the posterior tibial vein. The anterior and posterior tibial veins join below the knee to form the **popliteal vein** which continues as the femoral vein of the thigh and as the external iliac vein of the pelvis.

Blood from most abdominal organs does not drain into the inferior vena cava but into the hepatic portal system. The **hepatic portal system** receives blood from the pancreas, spleen, stomach, intestine, and gall bladder. The hepatic portal system consists of two capillary networks, one beginning in the digestive organs and one ending in the liver. Thus, the hepatic portal system delivers nutrient-rich blood to the liver for nutrient storage and metabolic conversion.

VEINS OF THE HEPATIC PORTAL SYSTEM

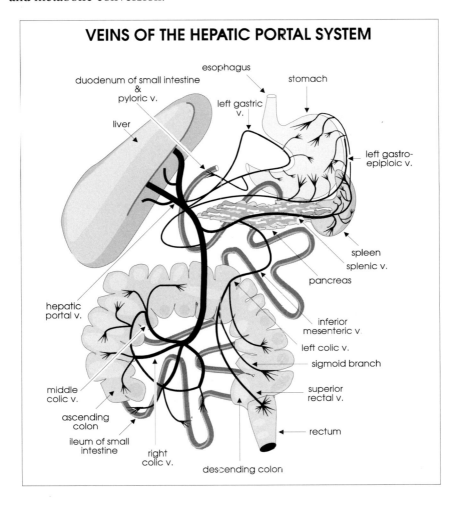

The **hepatic portal vein** is formed by the union of the splenic and superior mesenteric veins. The **splenic vein** receives blood from the pancreas, stomach, and portions of the large intestine while the **superior mesenteric vein** receives blood from the small intestine and portions of the stomach, large intestine, and pancreas. The **inferior mesenteric vein** drains the distal portion of the large intestine and empties into the splenic vein. The **gastric vein** from the stomach empties directly into the hepatic portal vein. Blood flowing through the liver empties into the **hepatic veins** which then empty into the inferior vena cava.

Fetal circulation is, for the most part, like that of the postnatal individual, but differs because the lungs, kidneys, and gastrointestinal organs do not function in the fetus. The fetus receives oxygen and nutrients by diffusion from maternal blood and eliminates carbon dioxide and waste products by diffusion into maternal blood. These exchanges occur through the **placenta**. The placenta is attached to the **umbilicus**, or navel, of the fetus by the umbilical cord which contains blood vessels. Blood is carried from the fetus to the placenta by two **umbilical arteries** which are branches of the internal iliac arteries. The umbilical arteries carry deoxygenated blood containing carbon dioxide and metabolic wastes from the fetus to the placenta for dumping by diffusion into the maternal blood. In the placenta, oxygen and nutrients diffuse from maternal blood into fetal blood which is then returned by the **umbilical vein** to the liver of the fetus.

The umbilical vein has two branches, one supplying the liver and one, the **ductus venosus**, that empties into the inferior vena cava, bypassing the liver. Consequently, oxygenated blood from the ductus venosus and deoxygenated blood are mixed before entering the right atrium of the heart. But, most blood bypasses the right ventricle in favor of the left atrium because of an opening in the septum between the right and left atria of the heart, the **foramen ovale**. Thus, most of the blood bypasses the pulmonary circuit and the nonfunctional, collapsed lungs. The blood that goes into the right ventricle is pumped into the pulmonary trunk, but most of this is then diverted into the short **ductus arteriosus** that connects with the aorta. The small amount of blood that does enter the pulmonary circuit nourishes lung tissues.

At or shortly after birth the lungs, kidneys, and digestive organs begin functioning. The umbilical arteries fill with connective tissue and become the **medial umbilical ligaments**. The umbilical vein becomes the **ligamen-**

tum venosum, or round ligament, of the liver and the ductus arteriosus becomes the **ligamentum arteriosum**. The **foramen ovale** closes, forming the **fossa ovalis**.

Chapter 5

THE LYMPHATIC AND DEFENSE SYSTEMS

The lymphatic system is part of the circulatory system that collects fluid leaking from capillaries for return to the main circulation. However, it also functions in defense against invading microorganisms and disease. **The lymphatic system** consists of a network of variously-sized vessels, lymphatics. The lymphatic capillaries, the smallest vessels, are blind-ended sacs. Lymphatic capillaries occur throughout the body with the exceptions of bone marrow, avascular tissue, and the central nervous system. Capillary walls permit interstitial fluid to flow in but not out because the simple squamous endothelial cells of the walls overlap to form one-way valves. Lymphatic capillaries lie in spaces between cells, draining excess interstitial fluid, or **lymph**. Higher fluid pressure outside capillaries causes endothelial cells to separate, permitting fluid entry.

The lymphatic system also transports absorbed dietary lipids to venous blood. Fat and fat-soluble vitamins are absorbed by lacteals, specialized lymphatic capillaries located in the center of villi of the mucosa that lines the small intestine. This lymph, called **chyle**, has a milky appearance.

Capillaries anastomose, or unite, forming larger lymphatic vessels that resemble veins but have thinner walls and more valves. Lymphatic vessels form in peripheral tissues and join together to form lymphatic trunks. Lymph of a pale golden color is returned to the venous system through two large thoracic-region ducts, the right lymphatic duct and the thoracic duct. The smaller **right lymphatic duct** receives lymph from the right upper extremity, head, and thorax. The larger **thoracic duct** arises from a sac-like collecting chamber, the **cycterna chyli**, and receives lymph from below the diaphragm and from the left side above the diaphragm. Both ducts empty at the junctions of the internal jugular and subclavian veins. The one-way transport of lymph is toward the heart. Because lymph is not pumped, transport depends upon pressure exerted by skeletal muscles and on pressure associated with breathing.

Lymph nodes are specialized, bean-shaped, lymph-filtering organs along lymphatic vessels. There are hundreds of lymph nodes with concentrations in the axillary, inguinal, and cervical regions. Lymph nodes are dense masses of lymphocytes and phagocytic macrophages covered by a capsule of dense connective tissue from which there extends collagenous trabeculae that divide the node into sinuses. A surface indentation, the hilus, is the point of attachment for blood vessels and nerves. Afferent lymphatic vessels convey lymph to the lymph node and efferent lymphatic vessels, attached at the hilus, carry lymph away, toward the venous system.

Lymphatic organs are the thymus gland, red bone marrow in flat bones and the epiphyses of long bones, the spleen, and clusters of lymphatic tissue like the tonsils. Lymphatic organs are clusters of lymphocytes, macrophages, and other cell types within a mesh of reticular connective tissue consisting of reticular cells and reticular fibers that are a fine form of collagen.

Red bone marrow is the only site of production of undifferentiated lymphoid stem cells that give rise to immune system T-cells, B-cells, and natural killer cells. As they mature, B-cells and natural killer cells enter the blood system. B-cells migrate to lymph nodes, the spleen, and other lymphoid tissues, while natural killer cells migrate throughout the body. Other lymphoid stem cells migrate to the thymus where they become T-cells.

The **thymus gland** is located in the thorax, anterior to the large vessels of the heart. A connective tissue capsule divides the thymus into two lobes which are themselves divided into lobules by fibrous partitions. Each lobule has a densely packed outer cortex and central medulla. The cortex is a mesh of epithelial reticular cells within which there are clusters of lymphocytes. The thymus is the site of maturation and processing of lymphocytes called **T-cells**. The hormone, **thymosin**, which is secreted by cortex cells, stimulates stem cell division and T-cell differentiation. But, while in the thymus, T-cells have no defensive function. Maturing T-cells move from the cortex to the medulla from which they exit via blood vessels. T-cells migrate to specific lymph nodes and the spleen, providing a defense against disease. The thymus gland attains maximum size at puberty then atrophies.

The **spleen**, lying along the curved border of the stomach, contains the most lymphoid tissue. The spleen filters blood much as do lymph nodes. The spleen is covered by a dense connective tissue containing collagen and elastic fibers. Trabeculae from the connective tissue capsule partition the spleen into sinuses. The white pulp of the spleen is lymphatic tissue, primarily B-cells, while the red pulp is blood-filled venous sinuses and thin cords of red blood cells, macrophages, lymphocytes, and plasma cells. Veins and lymphatic vessels exit at the hilum on the medial surface.

Within the red pulp, older red blood cells and platelets are destroyed and bacteria are phagocytized by lymphocytes produced in the white pulp. The spleen is also a reservoir for blood which is stored in sinusoids of red pulp. During fetal development, the spleen also produces blood cells.

Tonsils are incompletely encapsulated lymphatic tissues of which there are three groups beneath the pharyngeal epithelium: (1) the familiar palatine tonsils, (2) pharyngeal tonsils, and (3) lingual tonsils. The paired palatine tonsils lie in the lateral walls of the oropharynx. The pharyngeal tonsils, or adenoids, are in the nasopharynx while the lingual tonsils are at the base of the tongue. Tonsilar lymphocytes and macrophages protect against pathogens and harmful substances that enter at the nose and mouth.

Other clusters of lymphatic tissue are **Peyer's patches** in the mucous membrane of the gastrointestinal tract and the **vermiform appendix**, near the junction of the small and large intestines.

The body is continually exposed to pathogens and harmful environmental chemicals that may disrupt homeostasis and cause disease. **Resis-**

tance is the ability to counteract the potentially harmful effects through defense mechanisms. **Nonspecific defense mechanisms** prevent or limit the access and spread of microorganisms or hazardous chemicals. They include physical barriers, antimicrobial substances, complement proteins, phagocytes, inflammation, immunological surveillance, and fever.

Physical barriers are the first line of defense. The **skin** contains much **keratin**, a tough, water-proof protein, that provides a barrier to invading organisms. Hair and eyelashes provide protection from mechanical abrasion, dust, and insects. Body fluids like mucus, sweat, tears, saliva, and urine also provide nonspecific defenses. **Mucus**, which is secreted by mucus membranes, captures particles and invading organisms. Mucous membranes line cavities, like the respiratory and digestive tracts, that open to the outside environment.

Urine, tears, and saliva flush materials and organisms out of the body. These and other secretions, like **sebum** from sebaceous glands, contain antibacterial products such as **lysozymes**, providing a chemical barrier. Hydrochloric acid in the stomach is another chemical barrier, inhibiting the growth of most bacteria and degrading many otherwise harmful compounds.

Antimicrobial chemicals like transferrins, interferons, and complement proteins defend against the growth of microbes penetrating the skin and mucous membranes. **Transferrins**, found in blood and interstitial fluids, reduce the availability of iron to bacteria. **Interferons** are produced by virus-infected lymphocytes, macrophages, and fibroblasts. Interferons diffuse to uninfected neighboring cells and bind to cell surface receptors. This induces the synthesis of proteins that block viral reproduction. Interferon type II, or gamma interferon, also enhances the cell-killing activities of natural killer cells and phagocytic cells.

The complement system is blood plasma proteins that, when activated, enhance specific immune, allergic, and inflammatory responses. Most complement proteins are present as enzyme precursors, or proenzymes, until they contact a foreign substance. This triggers a chain of reactions resulting in the activation of all the proteins in the complement system. Activation of the complement system stimulates the release of histamine, the attraction of phagocytes, and the enhancement of phagocytosis.

Phagocytosis is the ingestion and destruction of microbes, cellular debris, or foreign compounds by phagocytes, microphages, and macrophages. **The primary microphages** are: (1) granulocytes or granular leukocytes, (2) neutrophils, and (3) eosinophils. **Neutrophils** are the first to arrive at a site of infection or tissue damage but they die after engulfing only a few bacteria. **Pus** is the accumulation of dead neutrophils, bacteria, and cellular debris. **Eosinophils** participate in allergic responses and also in the destruction of parasitic worms. **Monocytes** are chemotaxic white blood cells that migrate through blood plasma to infected, inflamed areas where they develop into larger **wandering macrophages**. In contrast, **fixed macrophages** reside in lymph nodes, the spleen, and other organs and tissues.

Phagocytosis requires binding between surfaces of the phagocyte and the microbe. Because a bacterium is surrounded by a polysaccharide capsule, it resists binding until coated, or **opsonized**, by antibodies or certain complement protein fragments. The coating proteins then interact with receptors on the phagocytes, triggering endocytosis.

Inflammation is a local, nonspecific defensive response to tissue damage caused by the penetration of the skin or mucous membranes by foreign particles or microbes. The redness and warmth associated with inflammation results from increased metabolic activity and blood flow in the damaged area. **Phagocytes** engulf the microbes or damaged cells, destroying them. **Mast cells** in the damaged tissues and surrounding connective tissue release **histamine** and other chemical mediators, such as **prostaglandins** and **leukotrenes**. The chemical mediators diffuse to nearby blood capillaries and arterioles, causing vasodilation and, therefore, increased blood flow. Diffusing chemicals also attract chemotaxic neutrophils and macrophages. Chemotaxis is movement in response to a chemical gradient. **Diapedesis** is the movement of neutrophils into tissue spaces by squeezing through capillary walls which have been rendered highly permeable by chemical mediators. Clotting factors form a mesh of fibrin that isolates the infected area, reduces blood loss and prevents the spread of bacteria.

Fever is abnormally high body temperature during infection and inflammation which may intensify the effect of antimicrobial substances, inhibit growth of some microbes, and accelerate repair processes.

Immunological surveillance is a nonspecific defense in which **natural killer cells**, a type of lymphocyte, destroy abnormal body cells as well as a variety of microbes by cytolysis. Here, the protein, **perforin**, perforates the plasma membrane, causing the cell to rupture. Natural killer cells occur in the spleen, red bone marrow, lymph nodes, and blood.

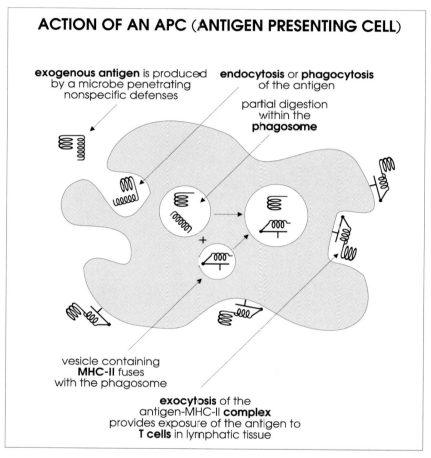

ACTION OF AN APC (ANTIGEN PRESENTING CELL)

exogenous antigen is produced by a microbe penetrating nonspecific defenses

endocytosis or **phagocytosis** of the antigen

partial digestion within the **phagosome**

vesicle containing **MHC-II** fuses with the phagosome

exocytosis of the antigen-MHC-II **complex** provides exposure of the antigen to **T cells** in lymphatic tissue

Nonspecific defense mechanisms provide immediate protection against a diverse array of pathogens and foreign substances. In contrast, **specific defense mechanisms** are targeted against specific bacteria, viruses, toxins, and other foreign tissues. **The immune system** remembers foreign agents and will launch attacks if they reenter the body. Specificity coupled with memory is the basis of immunity. A critical feature of the immune system is the ability of its cells to distinguish the body and its products from foreign organisms and their products.

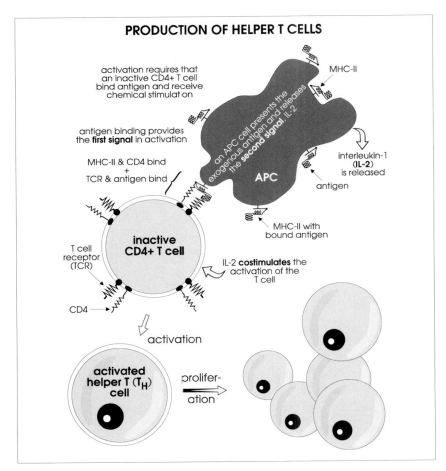

PRODUCTION OF HELPER T CELLS

activation requires that an inactive CD4+ T cell bind antigen and receive chemical stimulat on

MHC-II

antigen binding provides the **first signal** in activation

an APC cell presents the exogenous antigen and releases the **second signal**, IL-2

MHC-II & CD4 bind
+
TCR & antigen bind

interleukin-1 (**IL-2**) is released

APC

antigen

T cell receptor (TCR)

inactive CD4+ T cell

MHC-II with bound antigen

IL-2 **costimulates** the activation of the T cell

CD4 →

activation

activated helper T (T$_H$) cell

prolifer-ation

The primary cells of specific defense are **lymphocytes** and **macrophages**. Lymphocytes that develop immunocompetence are B-lymphocytes, or B-cells, and T-lymphocytes, or T-cells. T-cells become immunocompetent before leaving the thymus while B-cells become immunocompetent before leaving the red bone marrow. Immunocompetence entails acquiring special cell-surface antigen receptors that can bind specific antigens. Thousands of types of T-cells and B-cells are produced, each with a specific antigen receptor site. For instance, immunocompetent **T4** and **T8** cells have, respectively, **CD4+** or **CD8+ proteins** on their surface. About 70% of the circulating lymphocytes are T-cells. Most of the B-cells are localized in lymphoid tissues.

   Cell-mediated immunity is the killing of fungi, protozoan parasites, cancer cells, and virus-infected cells by T-cells. Foreign tissue cells may also be killed, resulting in transplant rejection. The immune response requires activation of T-cells through exposure to an antigen that is usually present in the glycoprotein matrix on the cell surface. These unique glycoproteins, called **major histocompatibility complex antigens**, occur on all cells except for red blood cells.

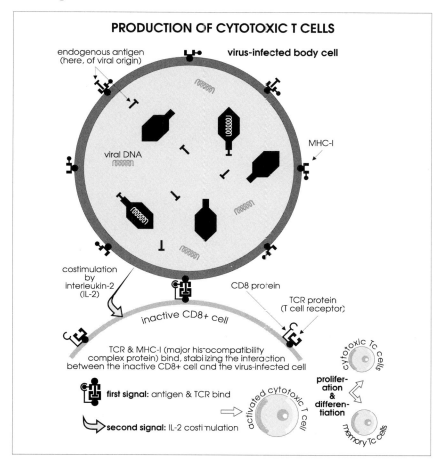

Antigen-presenting cells, including macrophages, dendritic cells, and B-cells, phagocytize the ultimate target cells, producing protein fragments which are then bound to major histocompatibility complex glycoproteins.

The antigen fragment and major histocompatibility complex glycoprotein are then exposed to the T-cell population.

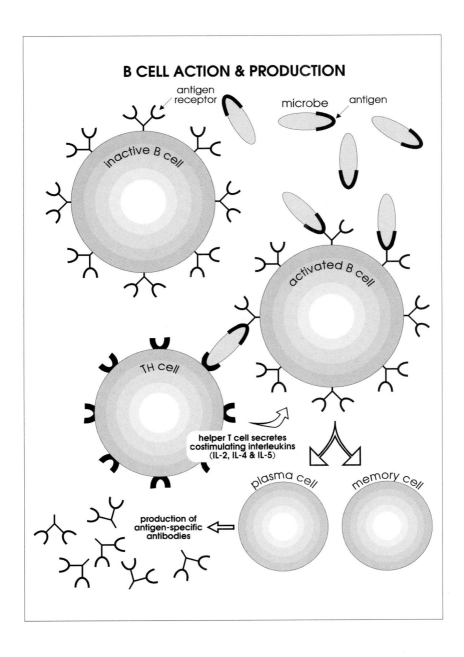

An immune response begins when the antigen is recognized by a T-cell with the appropriate antigen receptor site. T-cell proliferation and differentiation occurs when antigen recognition is coupled with the production of a chemical costimulator like internkin. **Activated T-cells** differentiate into four types of cells. One type, **helper T-cells**, secrete **cytokines** that destroy target cells and also stimulate B-cell production. Another type, **killer T-cells**, bind to target cells via antigen:receptor interaction and directly destroy them. **Memory T-cells** retain receptors for specific activating antigens, providing for a more intense response to subsequent encounter with the antigen. **Suppressor T-cells** depress the activity of other T- and B-cells, regulating the immune system.

B-cells cause antibody-mediated, or **humoral immunity**, responses. **Antibodies** are proteins that attach to the antigen to destroy, neutralize, or inhibit it. Antibodies are effective against bacteria, bacterial toxins, and viruses that occur outside body cells. Like T-cells, B-cells respond to the one specific antigen matching the antigen receptor site. B-cells can respond to unprocessed antigen in lymph or interstitial fluid, but their response is more intense if antigens are processed by macrophages.

B-cell activation begins when an antigen is bound to an antibody of the B-cell surface. The sensitized B-cell interacts with a helper T-cell that binds to the **major histocompatibility complex antigen** of the B-cell. In response to binding, the T-cell secretes cytokines that, in turn, promote B-cell proliferation and differentiation. B-cells differentiate into memory B-cells or into plasma cells that secrete millions of antibodies. The antibodies are then transported by blood and lymph. Antibodies are also called **immunoglobulins**, of which there are several classes designated as IgA, IgD, IgE, IgG, and IgM. IgG immunoglobulins, or **gamma globulins**, constitute 80-85% of the antibodies in blood.

Chapter 6

THE RESPIRATORY SYSTEM

The respiratory system delivers oxygen to the circulatory system and removes carbon dioxide. **The upper respiratory system** consists of the nose, pharynx, and associated structures, while the lower respiratory system consists of the larynx, trachea, bronchial tree, and lungs. As respiratory passages open to the outside, they are lined with a mucous membrane.

The **nose**, with its paired external nares or nostrils that open into the nasal cavity, is the primary passageway for air entering the respiratory system. The nasal mucosa warms and moistens incoming air while hairs within the nasal cavity trap dust and other foreign particles. The nasal cavity is the interior two chambers of the nose that are separated by a partition, the nasal septum. The **nasal septum** is composed in part of hyaline cartilage and in part from the fusion of the vomer, perpendicular plate of the ethmoid, the maxillae, and palatine bones. The nasal turbinates or conchae project from the walls of the nasal cavity and increase the surface area of the respiratory membrane.

SAGITTAL SECTION OF THE UPPER RESPIRATORY TRACT

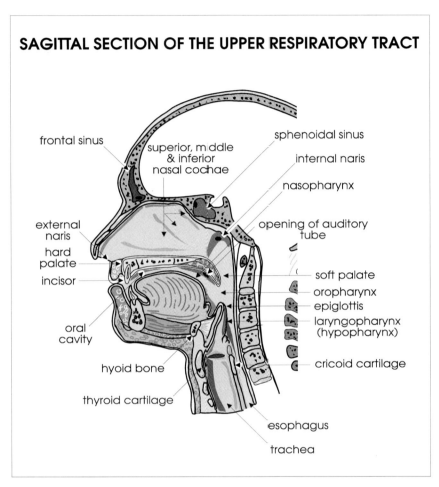

frontal sinus

superior, middle
& inferior
nasal conchae

sphenoidal sinus

internal naris

nasopharynx

external
naris

hard
palate

incisor

oral
cavity

opening of auditory
tube

soft palate

oropharynx

epiglottis

laryngopharynx
(hypopharynx)

cricoid cartilage

hyoid bone

thyroid cartilage

esophagus

trachea

As air moves around the **conchae** it is warmed by blood in capillaries. Also, mucus secreted by goblet cells moistens the air and traps dust particles and microorganisms. Cilia of the respiratory epithelium move the mucus to the pharynx where it is swallowed. Air flow is directed by three narrow troughs between the conchae: the superior, middle, and inferior meatus.

Paranasal sinuses are air-filled cavities in the frontal, maxillae, ethmoid, and sphenoid bones from which mucus drains into the nasal cavity. These secretions, along with tears from the nasolacrimal ducts, keep the nasal cavity moist and clean.

The **pharynx**, or throat, is a passageway from the internal nares to the larynx, connecting the nose and mouth. It has three regions: the naso-pharynx, oropharynx, and laryngopharynx.

The **nasopharynx** is superior to the soft palate, which separates it from the oral cavity. The auditory, or eustachian, tube opens into the naso-pharynx. The pharyngeal tonsils, or adenoids, are located on the posterior walls of the nasopharynx. The **oropharynx** is posterior to the oral cavity and extends from the soft palate to the level of the hyoid bone. The **fauces** is an opening from the mouth into the oropharynx. Within the oropharynx are the paired palatine and lingual tonsils at the back of the tongue. Below the oropharynx is the **laryngopharynx** which extends downward from the level of the hyoid bone, connecting the esophagus with the larynx. The oropharynx and the laryngopharynx are shared by the respiratory and di-gestive systems.

The **larynx**, or voice box, is the passageway between the pharynx, above, and the trachea, below. The larynx has walls formed by the large thyroid cartilage, cricoid cartilage, and epiglottis, and by the smaller, paired arytenoid, cuneiform, and corniculate cartilages. The thyroid cartilage, the largest, is connected to the hyoid bone by the ligamentous thyrohyoid mem-brane. The U-shaped **thyroid cartilage** forms the anterior and lateral walls of the larynx and represents the fusion of two plates of hyaline cartilage. The anterior surface of the thyroid cartilage is the "Adam's apple."

The thyroid cartilage rests on the ring-shaped cricoid cartilage to which it is connected by the cricothyroid ligament. The **cricoid cartilage** is the most inferior of the laryngeal cartilages. The **epiglottis** is a shield of elastic cartilage projecting above the glottis, the opening into the larynx. The narrow portion of the epiglottis is attached to the anterior rim of the thyroid cartilage while the broad portion is unattached and free to move up and down. During swallowing, the epiglottis covers the glottis, preventing the entrance of food.

The **glottis** consists of a pair of folds of mucous membrane, the vocal folds or true vocal cords, which are involved in sound production. The **rima glottidis** is the opening between the vocal folds.

The paired **arytenoid cartilages** are composed of hyaline cartilage. They are located on the superior and posterior border of the cricoid cartilage and attach to the vocal folds and to intrinsic pharyngeal muscles. The arytenoid cartilages influence the position and tension of the vocal folds.

The elastic, horn-shaped **corniculate cartilages** are a pair at the apices of the arytenoids. The paired **cuneiform cartilages** are club-shaped elastic cartilages that support the vocal folds and lateral epiglottis.

Sound waves are produced as air passes through the glottis, vibrating the vocal folds. The production of distinct words requires voluntary movements of the tongue, lips, and cheeks.

The **trachea**, or windpipe, is a 12 to 15 centimeter-long tube lined with ciliated pseudostratified columnar epithelium and mucus-secreting goblet cells. The trachea is anterior to the esophagus and extends into the mediastinum where it divides into right and left primary bronchi that enter the lungs.

There is a ridge of hyaline cartilage, the **carina**, at the bifurcation of the trachea into the primary bronchi. The **primary bronchi** divide to form smaller **secondary**, or **lobar, bronchi**. The right lung has three lobar bronchi while the left lung has two. The walls of the bronchi contain smooth muscle and cartilage but the amount of cartilage decreases as the bronchi get smaller. Asthma results from contraction of smooth muscles which constricts the air passages.

Secondary bronchi divide into tertiary, or segmental, bronchi, each of which supplies air to a single bronchopulmonary segment. **Bronchopulmonary segments** are divided into many smaller lobules each of which receives one terminal bronchiole. A **lobule** is a network of capillaries supported by elastic connective tissue and surrounding an alveolus. Tertiary bronchi ramify into **bronchioles**. Most bronchioles are lined with ciliated columnar epithelial cells and mucus-secreting goblet cells. However, the smallest terminal bronchioles lack goblet cells. Terminal bronchioles branch into smaller respiratory bronchioles that lead into microscopic alveolar ducts. Alveolar ducts end in clusters of air sacs, the **alveoli**, which are the sites of gas exchange.

Alveoli are lined with simple squamous epithelium containing macrophages and septal cells. Macrophages phagocytize any material reaching the alveolar surface. **Septal**, or surfactant, **cells** produce an oily secretion, surfactant, that reduces surface tension in the liquid coating the alveolar surface.

The **respiratory membrane** is a composite of the alveolar epithelium, its basement membrane, the interstitial space, another basement membrane, and the endothelial cells of the capillaries. Less than one micron separates alveolar air from blood, permitting rapid diffusion of oxygen and carbon dioxide across the respiratory membrane.

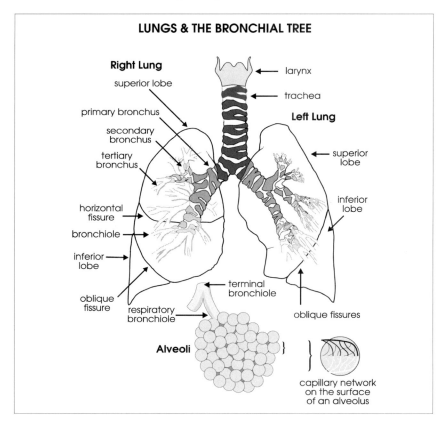

The cone-shaped pair of **lungs** lie in the thoracic cavity, resting on the diaphragm. The **mediastinum** is a partition between the air-filled, spongy lungs that extends from the sternum anteriorly to the vertebrae posteriorly and contains the pericardial cavity and heart. The apex of each lung is the narrow superior portion above the level of the first rib while the base is the concave inferior portion that rests on the superior surface of the diaphragm. The **hilus** is a depression on the posterior, medial surface of each lung. The hilus is where primary bronchi, pulmonary blood vessels, lymphatic vessels, and nerves enter and exit. The hilus is supported by the root, a mesh of dense connective tissue.

The right lung is shorter, wider, and has a greater air capacity than the left lung which has a medial concavity, the cardiac notch, where the heart lies. The right lung has three lobes that are separated by fissures. The horizontal fissure is between the superior and middle lobes while the oblique fissure is between the middle and inferior lobes. The left lung has one fissure, the oblique, separating the superior and inferior lobes. Each lobe is supplied by a secondary bronchus that branches into tertiary bronchi that supply bronchopulmonary segments, or lobules, that are defined by connective tissue septa. There are ten pyramid-shaped bronchopulmonary segments in each lung.

Each lung is enclosed in a double-layered serous membrane, the pleura or **pleural membrane**. The outer, parietal pleura attaches to the thoracic cavity wall while the inner, visceral pleura covers the lungs. Lubricating serous fluid is contained within the small pleural cavity that lies between the parietal and visceral pleura. Pleuritis, or pleurisy, is inflammation of the pleura.

Pulmonary ventilation, or breathing, is the sucking of air in, inspiration, and the forcing of air out, expiration. Air moves in and out because of gas pressure differences between the atmosphere and the inside of the lungs in accordance with **Boyle's law**. As gas enters a container, like a lung, it expands to fill it. As the volume occupied by a given amount of gas expands, the gas pressure decreases. Air, like all gases, flows from a region of higher pressure to a region of lower pressure. Thus, for inward air flow to occur, the lungs must expand, reducing the gas pressure within them. For expiration to occur, the lungs must contract, increasing the gas pressure within them.

Increase in size of the thoracic cavity accounts for expansion of the lungs because the outer visceral pleura of the lungs sticks to the parietal pleura lining the inner wall of the thoracic cavity. The primary inspiratory muscles are the diaphragm and external intercostals. Contraction of the diaphragm increases the length of the thoracic cavity. Contraction of the external intercostals pulls the ribs upward and pushes the sternum forward, increasing the width of the thoracic cavity. The sternocleidomastoid, scalenes, and pectoralis minor muscles assist with labored breathing.

Between breaths, **alveolar pressure** equals atmospheric pressure, 760 mm Hg at sea level. However, the pressure in the intrapleural space is about 2 mm Hg less than atmospheric pressure. The pressure gradient is

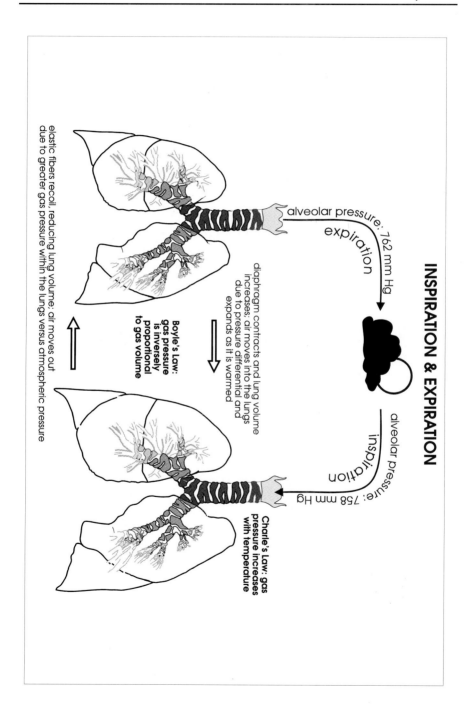

INSPIRATION & EXPIRATION

alveolar pressure: 762 mm Hg

expiration

elastic fibers recoil, reducing lung volume; air moves out due to greater gas pressure within the lungs versus atmospheric pressure

Boyle's Law: gas pressure is inversely proportional to gas volume

diaphragm contracts and lung volume increases; air moves into the lungs due to pressure differential and expands as it is warmed

alveolar pressure: 758 mm Hg

inspiration

Charle's Law: gas pressure increases with temperature

even greater during inspiration when the expanding rib cage enlarges the intrapleural space, lowering the intrapleural pressure. This causes the alveolar pressure to drop, but by a lesser amount. The pressure gradient between alveolar and intrapleural spaces prevents the lungs from collapsing. When the chest expands and the alveolar pressure drops below 760 mm Hg, the lower pressure draws air into the lungs. Newly inhaled air is mixed with air in the lungs. Consequently, oxygen and carbon dioxide concentrations in the lungs do not fluctuate greatly.

During expiration, the diaphragm and external intercostal muscles relax and elastic lung tissues recoil to their resting size. This results in alveolar pressure that is greater than atmospheric pressure, causing the bulk flow of air out of the lungs. At rest, expiration is passive, requiring no muscular contraction, but during exercise the internal intercostals and abdominal muscles actively expel air.

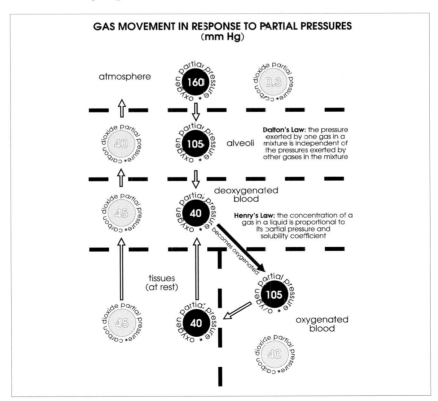

GAS MOVEMENT IN RESPONSE TO PARTIAL PRESSURES (mm Hg)

Dalton's Law: the pressure exerted by one gas in a mixture is independent of the pressures exerted by other gases in the mixture

Henry's Law: the concentration of a gas in a liquid is proportional to its partial pressure and solubility coefficient

The **resting breathing rate** is 12 to 15 inhalations and exhalations per minute. **Tidal volume** is the amount of air that enters and exits the lungs during a normal inspiration, about 500 milliliters. **Inspiratory reserve volume** is that amount in excess of tidal volume that is moved during a forced, deep breath. Inspiratory reserve volume may be as much as 3100 milliliters. The **expiratory reserve volume** is the amount expired in excess of tidal volume during forced expiration. It may be as much as 1200 milliliters. **Residual volume**, about 1200 milliliters, is the amount of air remaining in the lungs after the most forceful expiration. The residual volume prevents lung collapse.

Vital capacity is the sum of tidal, inspiratory reserve, and expiratory reserve volumes. Vital capacity is the maximum amount of air that can be expelled after a deep breath; typically about 4800 milliliters. **Total lung capacity**, about 6000 milliliters, is the sum of vital capacity and residual lung volume. **Inspiratory capacity**, about 3600 milliliters, is the sum of tidal and inspiratory reserve volumes. Inspiratory capacity is the maximum amount of air that can be inhaled.

External respiration is the diffusion of oxygen and carbon dioxide between alveoli and blood. **Internal respiration** is the diffusion of oxygen and carbon dioxide between blood and tissues. External and internal respiration reflect Dalton's and Henry's laws. **Dalton's law** is that the total pressure exerted by a mix of gases equals the sum of the pressures exerted by each gas independently. That is, total pressure is the sum of partial pressures. Air is about 79% nitrogen, 21% oxygen, four one-hundredths % carbon dioxide, and 46 one-hundredths % water vapor. Because air is 21% oxygen, the partial pressure of oxygen is 21% of 760 mm Hg or 159.6.

Henry's law is that each gas within a mixture of gases dissolves in a liquid in direct proportion to its solubility coefficient and partial pressure. Of the gases in air, the solubility in water is greatest for carbon dioxide, intermediate for oxygen, and least for nitrogen. The partial pressure of oxygen in alveoli, 105 mm Hg, is greater than in blood, 100 mm Hg. Therefore, the concentration gradient favors the diffusion of oxygen from alveoli into pulmonary capillaries. However, CO_2 diffuses in the opposite direction because its partial pressure in blood, 45 mm Hg is greater than that in alveoli, 40 mm Hg.

The rate of diffusion also depends on the surface area and thickness of the respiratory membrane. Diseases such as emphysema destroy alveolar walls, reducing the surface area for gas exchange. Ultimately, this limits the amount of oxygen reaching other tissues.

Oxygen first diffuses into the plasma and then into red blood cells where it combines with iron in hemoglobin molecules, forming oxyhemoglobin. About 98.5% of the oxygen diffusing into plasma is eventually combined with hemoglobin. The partial pressure of oxygen in blood, 100 mm Hg, is greater than that in tissues, 40 mm Hg. Consequently, 25% of the hemoglobin-bound oxygen diffuses from blood into tissues. Even more oxygen is released if the partial pressure of carbon dioxide in tissues is high, if the temperature is high, or if the pH is more acidic. Maternal blood entering the placenta transfers oxygen to fetal blood because of the greater oxygen carrying capacity of fetal hemoglobin.

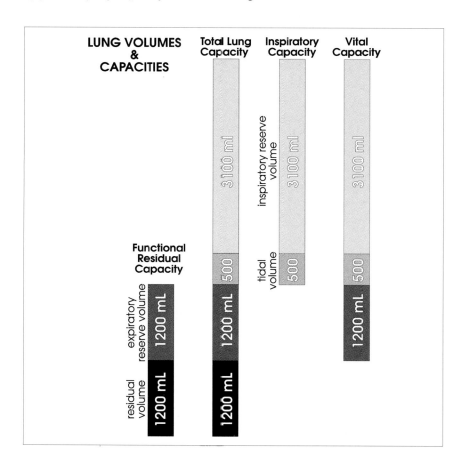

Cells require oxygen as the ultimate electron receptor in aerobic respiration and produce carbon dioxide as a by-product of catabolism. Consequently, the concentration of carbon dioxide in tissues is higher than in blood. Carbon dioxide diffuses out of tissues, into capillaries, and then into red blood cells where it is combined with water, producing carbonic acid (H_2CO_3). Carbonic acid dissociates into hydrogen ions that combine with hemoglobin and bicarbonate ions (HCO_3^-) that diffuse into the plasma. Bicarbonate ions account for approximately 70% of the carbon dioxide initially diffusing into blood while 23% is combined with hemoglobin. The last 7% remains dissolved in blood plasma.

Chapter 7

THE URINARY SYSTEM

Maintenance of body fluid homeostasis is a critical function of the urinary system. An adult body contains about 40 liters of fluid. Two-thirds of this is intracellular. One-fourth is interstitial fluid located primarily in the spaces between cells. Some interstitial fluid is localized in specific places: lymph in lymphatic vessels; cerebrospinal fluid in the brain; synovial fluid in joints; aqueous humor and vitreous humor in the eyes; endolymph and perilymph in the ears; pleural, pericardial, and peritoneal fluids between serous membranes; and glomerular filtrate in the kidneys. Blood plasma, the fluid portion of blood, accounts for the remaining 10% of the body fluid volume.

The **urinary system** filters the blood, ridding the body of metabolic wastes and helping maintain proper fluid and electrolyte balances. The urinary system consists of two kidneys, two ureters, one urinary bladder, and one urethra. The kidneys produce urine that passes through a pair of ureters to the urinary bladder where it is stored. Muscular contraction of the urinary bladder forces urine through the urethra from which it exits the body.

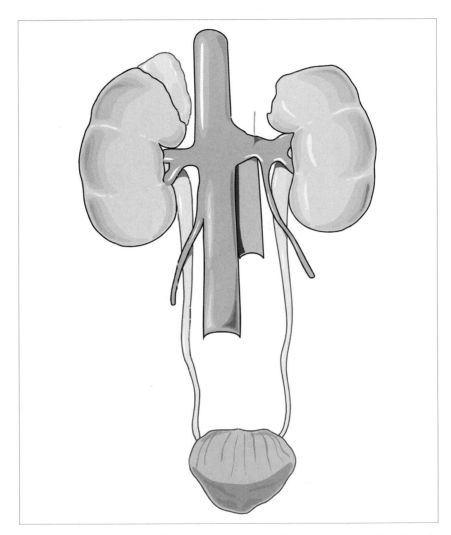

The paired **kidneys** lie between the twelfth thoracic and third lumbar vertebrae. But, the left kidney lies slightly higher than the right kidney. The kidneys are retroperitoneal organs, being located between the parietal peritoneum and the posterior wall of the abdomen. The kidneys are held in position by three layers of tissue. The outermost is the **renal fascia**, consisting of a thin layer of dense connective tissue that anchors the kidney to the back of the abdominal wall. The **adipose capsule**, the middle layer, consists of a mass of fatty tissue surrounding the renal capsule which is the innermost layer. The **renal capsule** is a smooth, transparent, fibrous membrane that is continuous with the outer membrane of the ureter at the hilus.

CORONAL SECTION OF A KIDNEY

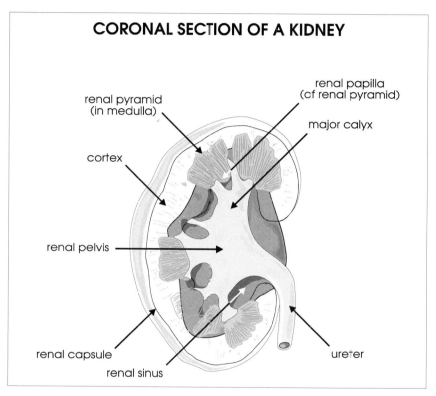

The **hilus** is a notch in the concave border of each kidney where a ureter is attached and where the renal artery, renal vein, lymphatic vessels, and renal nerves enter and exit. The hilus opens into a large cavity, the **renal sinus**. The **renal pelvis** is a funnel-shaped cavity inside the renal sinus that collects urine from nephrons and funnels it into the ureter. Emptying into the renal pelvis are 2 or 3 cup-shaped extensions called **major calyces**. Several smaller minor calyces join to form a major calyx. **Minor calyces** receive urine from collecting ducts of nephrons, the kidney's functional units.

The outer reddish-brown region of a kidney and its radial extensions, the renal columns, constitute the **cortex**. Six to 18 cone-shaped **pyramids**, covered by the cortex and separated by the renal **columns**, constitute the **inner medulla**. The apex of each renal pyramid is a **renal papilla** projecting into the lumen of a minor calyx that collects all the urine generated within the pyramid.

There are about one million nephrons per kidney. Each **nephron** consists of a renal corpuscle and a renal tubule. The **renal corpuscle**, which filters blood plasma, consists of a glomerulus and a glomerular capsule. The **glomerulus** is a cluster of filtering capillaries. The glomerular capsule, or **Bowman's capsule**, is a two-layered cap surrounding the glomerulus. Water and solutes pass from the capillaries through the inner visceral layer of the capsule into a capsular or Bowman's space enclosed by the inner visceral and outer parietal layers of the capsule.

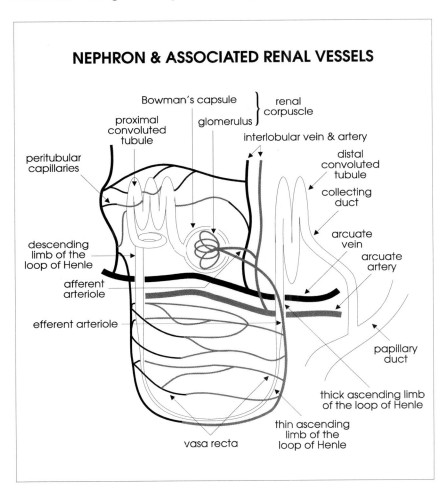

NEPHRON & ASSOCIATED RENAL VESSELS

The three-part inner visceral layer functions as a filter. It is composed of a glomerular endothelium of simple squamous cells that are fenestrated, or penetrated by many small pores. Blood plasma from the glomerular capillaries passes through the pores. A noncellular basement membrane that covers the endothelium prevents the passage of large proteins. Cells called **podocytes** cover the basement membrane. Numerous extensions of podocytes called pedicels form filtration slits that impede the passage of medium-sized proteins into Bowman's space. Thus, solute-rich plasma filtrate passes from the blood through the endothelial-capsular membrane into Bowman's space, producing urine.

Blood enters a glomerulus through an **afferent arteriole** and exits through an **efferent arteriole**. Filtration occurs as blood flows through the glomerular capillaries. Filtered fluid then passes from Bowman's space into the renal tubule.

The **renal tubule** has four main sections: (1) the coiled proximal convoluted tubule which is closest to the glomerulus, (2) the loop of Henle, (3) the distal convoluted tubule, and (4) the collecting duct which is farthest from the glomerulus. Cuboidal epithelial cells lining the proximal convoluted tubule have microvilli to reabsorb water and nutrients. These cells also secrete hydrogen and ammonium ions into the filtrate.

The U-shaped **loop of Henle** connects the proximal and distal convoluted tubules. Its **descending limb** dips into the medulla while its **ascending limb** extends upward into the cortex and back toward the glomerulus. At this point the tubule continues as the highly coiled **distal convoluted tubule**. As the ascending limb of the loop of Henle passes the glomerulus it contacts the afferent and possibly the efferent arteriole serving its glomerulus. Here, specialized cells in the tubule that monitor sodium chloride concentration are crowded together in the **macula densa**. The macula densa interacts with specialized smooth muscle cells in the arteriole that are called juxtaglomerular cells. With the macula densa, **juxtaglomerular cells** form a juxtaglomerular apparatus that contributes to the regulation of blood pressure and filtration rate.

The **juxtaglomerular apparatus** is an endocrine structure secreting the hormones, renin and erythropoietin. These hormones regulate blood pressure, blood cell production, and the rate at which the kidneys filter blood.

Several distal convoluted tubules empty urine into a single **collecting duct**. Many collecting ducts course through a renal pyramid and empty into about 30 larger papillary ducts. **Papillary ducts** widen at the renal papillae. The **renal papillae** empty into minor calyces of the renal pelvis.

Blood is brought to the kidneys by the right and left renal arteries that branch off of the abdominal aorta. Each renal artery divides into five segmental arteries as it enters a hilum. Each segmental artery branches into **interlobar arteries** that course through the renal columns between the renal pyramids. **Arcuate arteries** are branches of interlobar arteries that form arches between the bases of the renal pyramids in the outer cortex. Arcuate arteries branch into **interlobular arteries** that extend farther into the outer cortex where they divide into the afferent arterioles of the renal corpuscle.

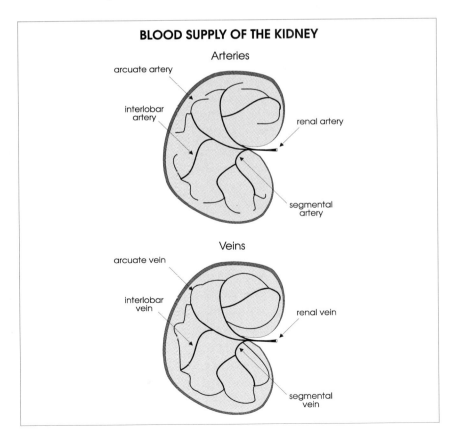

BLOOD SUPPLY OF THE KIDNEY

Afferent arterioles deliver blood to the glomerulus while efferent arterioles drain the glomerular capillaries of filtered blood. Each efferent arteriole divides into a network of peritubular capillaries that surround the proximal and distal convoluted tubules from which nutrients, ions, and water are returned to the blood.

There are two types of nephron. The glomerulus of a cortical nephron is in the outer portion of the cortex and a short loop of Henle just penetrates the outer medulla. 85% of nephrons are cortical nephrons. The glomerulus of a juxtamedullary nephron lies deep in the cortex and a long loop of Henle reaches through the medulla almost to the renal papilla.

In juxtamedullary nephrons, the peritubular capillaries connect to the vasa recta which are long capillaries associated with the loop of Henle. The peritubular capillaries and the vasa recta empty into a network of venules that then empty into the interlobular veins. The interlobular veins empty into arcuate veins that drain into interlobar veins running between renal pyramids. Interlobar veins drain into the segmental veins that in turn drain into the renal veins. Renal veins empty into the inferior vena cava.

Urine is formed as blood plasma is filtered during its passage through nephrons. First, glomerular filtration, which begins in the renal corpuscles of the nephrons, entails the flow of plasma from glomerular capillaries into nephrons. Glomerular filtration is driven by the hydrostatic, or blood, pressure in the glomerular capillaries. The filtration rate is proportional to the glomerular hydrostatic pressure minus the capsular hydrostatic pressure and osmotic force arising from the presence of plasma proteins in the blood. As blood plasma moves through pores in glomerular capillary endothelial cells and through filtration slits of podocyte, a filtrate of water and solutes passes into the capsular space. Large plasma proteins and blood cells do not pass through the filter. The glomerular filtration rate is approximately 1200 mL of blood passing through the kidneys per minute, producing about 125 mL of filtrate.

The initial filtrate leaves the capsular space, passing into the proximal convoluted tubule. About 99% of the useful substances dissolved within the filtrate are reabsorbed by the peritubular and vasa recta capillaries, primarily along the proximal convoluted tubule. Reabsorbed filtrate has high concentrations of nutrients such as glucose, amino acids, and lactic acid and of ions such as sodium. Reabsorption occurs by passive processes of simple diffusion, facilitated diffusion, and osmosis, and by active processes of active transport or, for peptides, pinocytosis. **Metabolic wastes** are not easily reabsorbed. They are: (1) **urea**, the most abundant organic waste, a toxic product of amino acid breakdown, (2) **creatinine** which is generated in skeletal muscle when creatine phosphate is metabolized, and (3) **uric acid** which arises from the recycling of nitrogenous bases found in the nucleic acids, RNA and DNA.

Microvilli of cells of the convoluted tubule greatly increase its surface area and account for 60 to 70% of **water reabsorption**. The loop of Henle accounts for 10 to 15% of reabsorption of water while the distal convoluted tubule and collecting duct account for 15 to 25%.

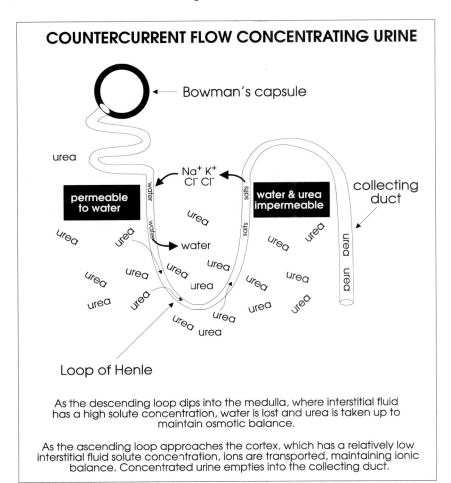

COUNTERCURRENT FLOW CONCENTRATING URINE

As the descending loop dips into the medulla, where interstitial fluid has a high solute concentration, water is lost and urea is taken up to maintain osmotic balance.

As the ascending loop approaches the cortex, which has a relatively low interstitial fluid solute concentration, ions are transported, maintaining ionic balance. Concentrated urine empties into the collecting duct.

Sodium ions diffuse from the filtrate through channels in the microvilli into tubule cells. Sodium is then actively pumped into the basolateral membrane and from there they diffuse into the interstitial fluid. Sodium ions diffuse from the interstitial fluid into peritubular capillaries. The active pumping of ions and other solutes in the proximal convoluted tubule produces osmotic flow of water into peritubular fluids, reducing the volume of the filtrate by 60 to 70% before it reaches the descending limb of the loop of Henle. Wastes are partially reabsorbed, including over 40% of the urea.

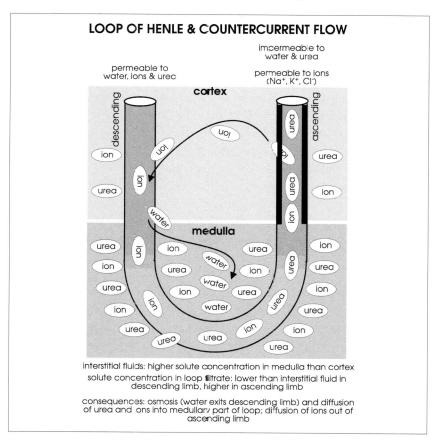

LOOP OF HENLE & COUNTERCURRENT FLOW

impermeable to water & urea

permeable to water, ions & urea

permeable to ions (Na+, K+, Cl-)

cortex

descending / ascending

medulla

interstitial fluids: higher solute concentration in medulla than cortex

solute concentration in loop filtrate: lower than interstitial fluid in descending limb, higher in ascending limb

consequences: osmosis (water exits descending limb) and diffusion of urea and ions into medullary part of loop; diffusion of ions out of ascending limb

In the loop of Henle, fluid in the descending limb loses water osmotically and gains urea by diffusion as it dips from the outer to inner medulla where solute concentrations in the interstitial fluid are high. Consequently, solute concentrations in the fluid rise. But, as the ascending limb passes back into the outer medulla its now relatively concentrated fluid can lose sodium and chloride ions to the interstitial fluid by diffusion. However, water is not regained osmotically and urea is not lost by diffusion because the ascending limb is impermeable to these. The process of water loss, urea gain, and ion loss in different parts of the U-shaped loop of Henle is called the **counter current mechanism**. Tubular cells in the ascending limb actively transport sodium and chloride ions back into the filtrate.

The distal convoluted tubule and collecting duct reabsorb sodium and chloride ions and water. Because these are the last regions to reabsorb metabolites, the osmotic concentration of urine is controlled in these parts of the renal tubules. In **principal cells** of the distal convoluted tubule, sodium-potassium exchange pumps and chloride leakage channels in the basolateral membrane transport ions into the interstitial fluid from which they move into peritubular capillaries. The hormone, **aldosterone**, acts to increase sodium reabsorption by principal cells. The presence of **antidiuretic hormone** (ADH) increases the water permeability of principal cells, increasing the reabsorption of water. Bicarbonate ions, formed in the **intercalated cells** by the breakdown of carbonic acid, are also reabsorbed.

When the plasma concentration of a substance is abnormally high, for instance, glucose in diabetes mellitus, there is not sufficient time for complete reabsorption and the substance appears in urine. The maximum rate of reabsorption for a substance is the transport maximum. Renal threshold is the plasma concentration at which a substance like glucose appears in urine.

Tubular secretion, another process in urine formation, entails removal of hydrogen, ammonium, and potassium ions, nitrogenous wastes such as urea, creatinine, and uric acid, and certain drugs like penicillin from the blood in the peritubular capillaries. Secretion of hydrogen ions occurs in epithelial cells of the proximal convoluted tubules and collecting ducts where carbon dioxide combines with water to form carbonic acid in a reaction catalyzed by **carbonic anhydrase**. Carbonic acid then dissociates into hydrogen ions and bicarbonate ions. The hydrogen ions are secreted into the tubular fluid while the weakly basic bicarbonate ions are reabsorbed into the blood. In this way, nephrons control blood pH.

Nephrons control blood concentration and volume by removing water and solutes, especially ions, from blood. Potassium ion secretion varies with dietary intake and is stimulated by the hormone, aldosterone, which acts on the principal cells of the distal tubules and collecting ducts. The secretion process involves the exchange of potassium for sodium ions.

Three hormones are secreted or activated in the kidneys: erythropoietin, calcitriol, and angiotensin II. **Erythropoietin** is secreted by the kidneys and stimulates the production of red blood cells in bone marrow, resulting in an increase in blood volume. **Calcitriol** is activated in the kidneys and stimulates the reabsorption of dietary calcium and phosphate in the gastrointestinal tract, thereby, increasing blood plasma concentrations. **Angiotensin II**, a powerful vasoconstrictor responsible for increasing blood pressure, is activated through the action of renin, an enzyme secreted by juxtaglomerular cells in the walls of afferent arterioles.

Renin is secreted when the sodium concentration in the macula densa is low, which occurs when blood pressure is low and when the rate of filtration in the kidneys is low. Juxtaglomerular cells are sensitive to pressure changes. When blood pressure is low in the afferent arteriole, renin is secreted.

The concentration and volume of urine is determined by the amount of water reabsorbed from the distal convoluted tubule and collecting duct and is regulated by three hormones: aldosterone, antidiuretic hormone (ADH), and atrial natriuretic hormone. Angiotensin II in the blood stimulates the adrenal cortex of the kidney to secrete aldosterone. Aldosterone promotes sodium reabsorption in the distal convoluted tubule and collecting duct. As sodium is reabsorbed, water follows by osmosis, reducing urine output. Angiotensin II also stimulates the secretion of antidiuretic hormone by the posterior pituitary gland. ADH stimulates reabsorption of water in the distal convoluted tubule and collecting duct. **Atrial natriuretic hormone**, or **atriopeptin**, is produced by specialized heart cells. Atrial natriuretic hormone promotes the excretion of sodium and water by inhibiting the secretion of antidiuretic hormone. Thus, atrial natriuretic hormone increases the amount of dilute urine produced, simultaneously decreasing blood pressure and volume.

The physical and chemical characteristics of urine are determined by changes in diet, physical activity, and health. Therefore, **urinalysis**, the chemical and microscopic examination of urine, reveals information about an individual's physiological condition. Daily urine production varies but averages 1 to 2 liters. The more concentrated the urine, the darker its color. The color is due to the pigment, **urochrome**, produced from the breakdown of bile. However, foods, drugs, and diseases may affect the color as well as the odor. The odor becomes ammonia-like upon standing but that of diabetics is sweet due to the presence of glucose and ketone bodies. The pH of urine averages about 6.0 but varies with diet, ranging between 5.0 and 7.8. High protein diets lower the pH while vegetarian diets raise the pH. The specific gravity (grams/ml) of urine ranges from 1.008 to 1.030 with the higher values corresponding to higher salt concentrations.

The chemical composition of urine is about 95% water and 5% solutes derived from cellular metabolism. Organic solutes consist primarily of urea, uric acid, and creatinine. **Urea** is produced when liver amino acids are deaminated, producing ammonia which is then combined with carbon dioxide. Urea constitutes 60 to 90% of the nitrogenous waste in urine. **Uric acid** is produced from the catabolism of DNA and RNA. Uric acid may crystallize, giving rise to kidney stones. **Creatinine** is derived primarily from the breakdown of creatine phosphate in muscle tissues. Other substances such as ketone bodies, carbohydrates, pigments, fatty acids, mucin, enzymes, and hormones may be present in minute quantities. **Ketone bodies** are produced during excessive triglyceride catabolism and appear in high concentration in individuals with diabetes mellitus or suffering acute starvation. Inorganic solutes include sodium, potassium, ammonium, magnesium, calcium, chloride, sulfate, and phosphate ions.

Urine is collected in the renal pelvis from which it flows into the **ureters**, a pair of muscular tubes, one from each kidney, extending about 30 centimeters to the urinary bladder. Transport of urine to the bladder is by **peristalsis**. The ureter walls consist of three layers. The **mucosa** is the inner layer of transitional epithelium. The **muscularis** is the middle layer of inner longitudinal and outer circular smooth muscle sublayers. The **adventitia** is the outer layer of fibrous connective tissue.

The hollow, muscular **urinary bladder** is a retroperitoneal organ lying in the pelvic cavity posterior to the pubic symphysis. The superior surface of the bladder is covered by the peritoneum. In males, the bladder is directly anterior to the rectum but in females it is anterior to the vagina and inferior to the uterus.

Three layers of tissue form the wall of the urinary bladder. The inner **mucosa** consists of transitional epithelium with **rugae** or folds that permit stretching. There is also underlying connective tissue, the **lamina propria**. Some cells secrete a mucus that protects the epithelium from urine. The middle **detrusor muscle layer** consists of three separate sublayers of smooth muscle, some fibers of which surround the opening to the urethra to form the internal urethral sphincter. The **adventitia** is an outer layer of fibrous connective tissue. The **trigone** is a triangular area of the bladder floor. The two ureteral openings are at the two posterior corners of the trigone. The **internal urethral orifice**, which opens to the urethra, is at the anterior corner of the trigone.

The **urethra** is a tube for the elimination of urine that leads from the floor of the urinary bladder. In males, the urethra also serves as a passageway for the discharge of semen. The urethra is 15 to 20 centimeters long in males but only 4 centimeters long in females. A modification of the urogenital diaphragm consisting of skeletal muscle forms the external urethral sphincter which is under voluntary control. The **external urethral orifice** is the opening of the urethra to the exterior. In males, the wall of the urethra is an inner mucosa, lamina propria, and adventitia .

In females the wall of the urethra consists of an inner mucosa, a thin middle layer of spongy tissue containing a plexus of veins, and an outer serosa that is continuous with the serosa of the urinary bladder. The **external urethral orifice** is located between the clitoris and the vaginal opening.

The urinary bladder can hold 700 to 800 mL of urine. However, when the volume of urine reaches 200 to 400 mL, the bladder wall is sufficiently distended to stimulate stretch receptors, initiating the micturition reflex. **Urination** or micturition is a reflex response involving both involuntary and voluntary nerve impulses. The response is coordinated in the **micturition reflex** center in the sacral region of the spinal cord. Impulses from the center travel via sensory nerve tracts to the cerebral cortex, producing a conscious desire to urinate. Impulses travel back to the urinary bladder via parasympathetic nerves, causing the detrusor muscle in the wall of the bladder to contract rhythmically and causing the internal urethral sphincter to relax.

Incontinence is a lack of voluntary control over micturition that occurs in infants less than two years of age or in adults as a consequence of either urinary bladder disease, damage to the external sphincter, or inability of the detrusor muscle to relax when emotionally stressed. **Retention** or failure to completely or normally void urine may be caused by an obstruction in either the urethra or neck of the urinary bladder, by nervous contraction of the urethra, or by lack of sensation to urinate.

CHAPTER 8

THE DIGESTIVE SYSTEM

The digestive system breaks down food into small molecules, which are then absorbed by the circulatory system for delivery to cells. The digestive system also eliminates undigested material from the body. The **digestive system** consists of the alimentary canal, or gastrointestinal (GI) tract, and the accessory structures. The GI tract is a continuous, highly modified tube with an opening at each end, the mouth and the anus. The regions of the GI tract are the oral cavity, pharynx, esophagus, stomach, small intestine, and large intestine. Accessory structures include the teeth and tongue, and organs like the liver and pancreas that empty their contents into the GI tract via ducts.

Digestion entails ingestion, mechanical and chemical decomposition of food, absorption, and elimination. Ingestion, or taking food into the oral cavity, is followed by mechanical breakdown of food by chewing or by contractions of the GI tract. Chemical digestion involves the breakdown of large molecules by enzymatic action. Absorption is the passage of digested molecules, primarily from the small intestine, into the blood and lymph capillaries. Feces, consisting of undigested food, blood, and bacteria, are defecated at the anus.

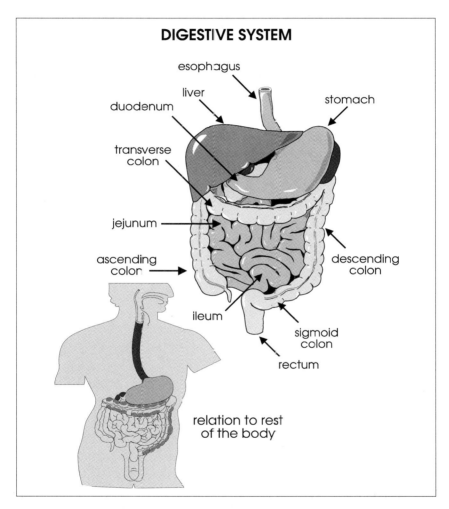

With the exception of the oral cavity, the wall of the digestive tract consists of four tissue layers, or tunics: the mucosa, submucosa, muscularis externa, and serosa. The **mucosa** lining of the digestive tract is a mucous membrane. Its inner surface is stratified squamous epithelium in areas of high abrasion like the mouth, esophagus, and anus and a simple, columnar epithelium elsewhere. Beneath the mucosa is the **lamina propria**, an areolar connective tissue containing blood and lymphatic vessels. The outer layer, the **muscularis mucosa**, has smooth muscle fibers that impart a ridged texture.

The **submucosa** is external to the mucosa, and consists of large blood and lymphatic vessels, submucosal glands, and a network of autonomic nerves called the submucosal plexus or the **plexus of Meissner**.

The **muscularis externa** is primarily an inner circular and outer longitudinal layer of smooth muscles. A network of autonomic nerve fibers, the myenteric plexus or the **plexus of Auerbach**, coordinates movements of the muscularis. **Peristalsis** is rhythmic contractions of the muscularis that propel contents through the GI tract.

The outer serosa is connective tissue above the diaphragm, the adventitia portion, and a single layer of serous epithelium, the serosa portion, in the abdominal cavity. The serosa portion constitutes the visceral peritoneum that covers organs of the GI tract.

The inner body wall is also lined with serous membrane, the parietal peritoneum. The two serous membranes secrete lubricating serous fluid into the peritoneal cavity between them. Peritonitis is an acute inflammation of the peritoneum and is often-life threatening. Some organs of the GI tract are suspended within the peritoneal cavity by double sheets of serous membrane, the mesenteries, that connect the parietal to the visceral peritoneum.

The mouth, or **oral cavity**, consists of the lips, cheeks, palate, and the accessory structures, the tongue and teeth. It is lined with stratified squamous epithelium. On the gums, hard palate, and dorsal surface of the tongue, the epithelium is slightly keratinized, providing protection against abrasion during eating. The lips, or labia, are folds of skeletal muscle covered by a thin layer of skin. They contain many sensory receptors for determining the temperature and texture of food. The **labial frenulum** is a layer of mucous membrane that attaches each lip to the gum at the midline. The gums, or **gingivae**, surround the base of each tooth on the alveolar surfaces of the maxilla and mandible. The **vestibule** is the space between the cheeks and lips, and the gums and teeth. The **fauces** is the opening connecting the oral cavity and the pharynx.

The palate, or roof of the mouth, separates the oral cavity from the nasal cavity. The **hard palate** is an anterior bony shelf while the posterior **soft palate** is muscle tissue covered by mucous membrane. The soft palate terminates in the dangling **uvula,** and two lateral folds, the muscular pharyngeal arches. The soft palate and associated structures direct food into the oropharynx.

The large, moveable tongue is composed of confined **intrinsic muscles** that are not attached to bone and **extrinsic muscles** that attach to other structures in the mouth. The underside of the tongue connects to the floor of the mouth by the sheet-like **lingual frenulum**. Posteriorly, the tongue attaches to the hyoid bone. The superior surface of the tongue is covered by chemosensory circumvallate and fungiform **papillae** or taste buds.

The **dentition** consists of an upper set of teeth located in the maxilla, and a lower or mandibular dentition. There are four tooth types: (1) slicing, shovel-like **incisors**, (2) piercing, cone-shaped **canines**, (3) grinding and slicing bicuspids, or **premolars**, and (4) grinding **molars**. The deciduous dentition (baby teeth) of two incisors, one canine, and two molars on each side of the upper and lower jaws is replaced in childhood. The permanent dentition is two incisors, one canine, two premolars, and three molars on each side of the upper and lower jaws, totaling 32 teeth. The third molars are the wisdom teeth.

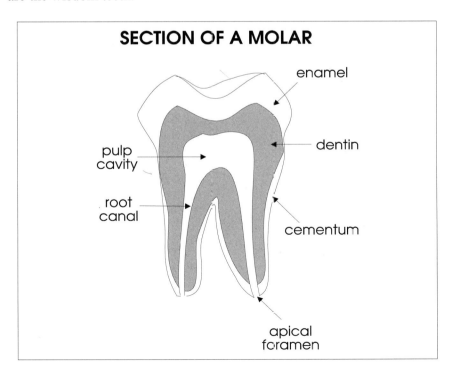

SECTION OF A MOLAR

enamel

pulp cavity

dentin

root canal

cementum

apical foramen

The crown is the visible portion of the tooth. Its outer layer is **enamel**, a hard, mineralized tissue consisting primarily of calcium phosphate. Deep to the enamel is **dentin**, a calcified connective tissue. **Caries**, or dental decay, is caused by demineralization due to the action of bacteria. An inner pulp cavity contains connective tissue, blood vessels, lymphatic vessels, and nerves.

The neck marks the junction of the crown with the roots, which fit into sockets in the jaw. A tooth may have one to three roots. **Roots** lack enamel but are covered by a mineralized bone-like substance, **cementum**. Roots contain pulp in root canals that are continuous with the pulp cavity. The base of each root canal is an apical foramen, through which blood and lymphatic vessels and nerves enter and exit. Tooth roots are held in sockets, or alveoli, by dense fibrous connective tissue bands, the periodontal ligaments.

Salivary glands produce and secrete saliva which is 99% water and 1% the digestive enzyme salivary amylase, plus ions, glycoproteins, lysozyme, and metabolic wastes. Saliva moistens and lubricates food particles, dissolves chemicals that stimulate the taste buds, and initiates digestion of complex carbohydrates. Most saliva is produced by three pairs of large extrinsic salivary glands, external to the mouth. The largest, the **parotid gland**, is anterior to the ear and external to the masseter muscle. The parotid duct enters the mouth near the second maxillary molar. Mumps is an inflammation of the parotid glands caused by a myxovirus. The **submandibular gland** is beneath the base of the tongue, along the medial aspect of the mandible. Its duct runs beneath the mucosa of the oral cavity floor, opening at the base of the lingual frenulum. The small **sublingual gland** is superior to submandibular gland.

During mastication, the food is mixed with saliva and shaped into a ball, the **bolus**, by the tongue and palate. **Salivary amylase** initiates chemical digestion of complex polysaccharides into disaccharides. The bolus is pushed into the oropharynx by the tongue. This voluntary swallowing is called **deglutition**.

The **pharynx** is a fibromuscular passageway about 12.5 centimeters long that connects the nasal and oral cavities to the larynx and esophagus. Three pharyngeal regions are: (1) the **nasopharynx**, running from the internal nares to the base of the soft palate, (2) the **oropharynx**, where the fauces of the oral cavity opens, and (3) the **laryngopharynx**, extending from the oropharynx at the level of the hyoid bone to the opening of the larynx and esophagus.

The **esophagus** is a collapsible muscular tube about 30 centimeters in length, running from the pharynx to the stomach. It passes through the diaphragm via an opening called the esophageal hiatus. Involuntary contractions of the muscularis layer of the pharynx and esophagus move the bolus. The mucosa secretes mucus that lubricates the bolus. Upper and lower esophageal sphincters are valves that control the movement of food into and out of the esophagus.

The **stomach** is an enlarged, sac-like region of the GI tract. The esophagus enters the cardiac region. The **fundus** is the most superior portion, extending above and to the left of the cardiac region. Inferior to the cardiac and fundic regions, the body of the stomach is J-shaped. The inner, or the lesser, curvature, is directed superiorly and to the right. The body narrows inferiorly to form the **pyloric region**, which empties into the small intestine. The portion of the pyloric region that is continuous with the body is known as the **pyloric antrum**, the region that contacts the small intestine is the **pyloric canal**. The **pyloric sphincter**, at the junction of the pyloric canal and duodenum of the small intestine, regulates the release of partially digested food, known as **chyme**, from the stomach.

The stomach muscularis has an inner oblique layer of smooth muscle in addition to the circular and longitudinal layers. The mucosa is folded into ridges, or **rugae**, which allow the stomach to expand as it fills. Microscopically, the mucosa is folded into **gastric pits**, narrow channels that extend into the lamina propria. At the base of these gastric pits are the **gastric glands**, which contain parietal cells, chief cells, mucus cells, and enteroendocrine cells. Secretions of these cells produce gastric juice. **Parietal cells** transport hydrogen and chloride ions into the lumen of the stomach, where they react to form hydrochloric acid. HCl kills pathogens in food, and also denatures proteins. Parietal cells also secrete intrinsic factor, a substance necessary for the absorption of vitamin B12. **Chief cells** secrete pepsinogen, the inactive form of the protein-digesting enzyme pepsin. Pepsinogen is converted into pepsin by the low pH of the gastric juices. **Mucus-secreting cells** are abundant in the cardiac region but are also found in other regions. A thick alkaline mucus coats and protects the stomach mucosa. A watery mucus aids in digestion. **Enteroendocrine**, or G, **cells**, located primarily in gastric pits in the pyloric region, release the hormone, gastrin. Gastrin stimulates a number of digestive processes in the stomach.

Gastric juice is produced continuously but the amount and composition is regulated by the central nervous system and hormones. **Cephalic phase** regulation begins when an individual encounters, or, sometimes, merely thinks about food. Parasympathetic innervation of the stomach via the vagus nerve is initiated by the medulla oblongata, resulting in increased stomach motility and secretion of **gastrin,** which triggers the release of gastric juice. Most gastric juice is produced during the **gastric phase** of regulation, when the presence of food in the stomach stimulates churning and secretion. As stomach pH falls to about 2, pepsinogen is converted to pepsin and proteins are hydrolyzed into smaller units. Low pH also inhibits further secretion of gastric juices by a negative feedback mechanism.

The release of acidic chyme, a mix of gastric secretions and partially digested food, into the small intestine triggers an intestinal phase of regulation. Distension of the small intestine by chyme stimulates both the neuronal enterogastric reflex and the release of intestinal hormones. The hormones, **gastric inhibitory peptide** and **secretin**, inhibit stomach motility and gastric juice secretions, while **cholecystokinin** influences the rate of chyme release into the small intestine.

Food is moved in the stomach by peristaltic contractions that produce mixing waves that, then, produce chyme. As chyme accumulates, the pyloric sphincter relaxes, and small amounts are pumped into the small intestine. The rate of chyme release depends on the nature of the food and the receptivity of the small intestine, but the stomach usually empties within four hours after a meal. Liquids pass through the stomach quickly, while solid materials remain until they are well-mixed with gastric juice. Carbohydrates, already partially digested by salivary amylase, pass through quickly. Proteins and fats may remain in the stomach for six hours.

The **small intestine** is about 6 meters in length. Its diameter ranges from about 4 centimeters at the pylorus to about 2.5 centimeters at its junction with the large intestine. The **duodenum** is the first 25 centimeters of the small intestine. It receives chyme from the stomach pylorus and digestive secretions from the pancreas and liver. Within the duodenum, submucosal glands produce large quantities of mucus to protect the epithelium from the acid chyme arriving from the stomach. They also secrete a hormone, **urogastrone**, which inhibits gastric acid production. The duodenum provides buffers that are mixed with chyme, elevating its pH from 2 to 7 or 8.

The **jejunum** is the middle section of the small intestine, about 2.5 meters in length. At the junction of the duodenum and jejunum, the small intestine enters the peritoneal cavity, and is supported by sheets of mesentery. The final stages of chemical digestion and much absorption occur in the jejunum.

The 3.5-meter long **ileum** is the last region of the small intestine, terminating at a sphincter, the **ileocecal valve**, which controls the flow of chyme from the ileum into the cecum of the large intestine. The ileum possesses some lymphoid tissues, **Peyer's patches**.

Localized contractions in the small intestine result in segmentation, which mixes intestinal juices and chyme. Large-scale contractions of the mucosa propel the chyme through the small intestine by peristalsis. Both segmentation and peristalsis are regulated by the autonomic nervous system.

The mucosa of the small intestine has transverse folds, the **plicae circulares**, which increase the surface area for absorption. The absorptive surface area of the mucosa is also enhanced by finger-like folds called villi. The simple columnar epithelial cells of the mucosa possess **microvilli**, hair-like extensions of the cell membrane that extend into the intestinal lumen. Collectively, the microvilli form the "brush border" of the mucosa.

The **villi** are the sites of absorption. The center of each villus consists of lamina propria that contains a blood capillary and a lymphatic capillary. The lymphatic capillary, or lacteal, receives nutrients, such as some fats, that are too large to diffuse into the blood capillaries.

At the base of the villi, the mucosa is studded with pits called intestinal crypts or **crypts of Lieberkühn**. Epithelial cells that line these pits secrete **intestinal juice**, a carrier fluid for the absorption of nutrients from chyme. These cells also produce and release the enzymes **maltase**, **sucrase**, and **lactase**, which reduce carbohydrates to simple sugars, or monosaccharides, **nucleases** which break down nucleic acids, and **peptidases** which reduce polypeptides to peptides and amino acids. Some digestive enzymes are secreted by the pancreas and gall bladder.

Mucus-secreting goblet cells and enteroendocrine cells are also scattered throughout the mucosa. Enteroendocrine cells secrete hormones. Vasoactive intestinal polypeptides stimulate the production of intestinal juice.

In addition to their inhibition of gastric activity, secretin and cholecystokinin also stimulate the release of pancreatic juice. Cholecystokinin also stimulates the gallbladder to release bile into the duodenum.

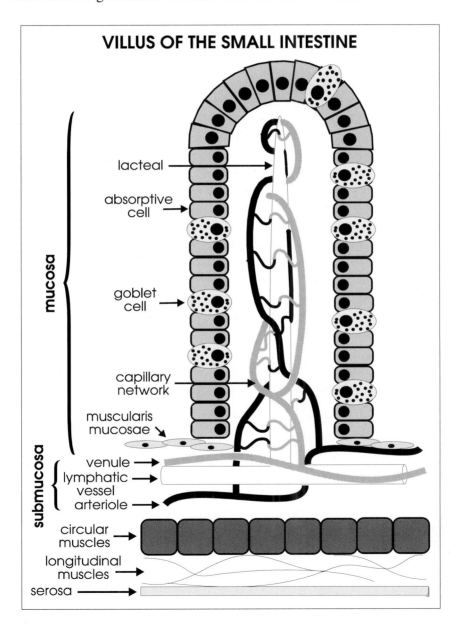

VILLUS OF THE SMALL INTESTINE

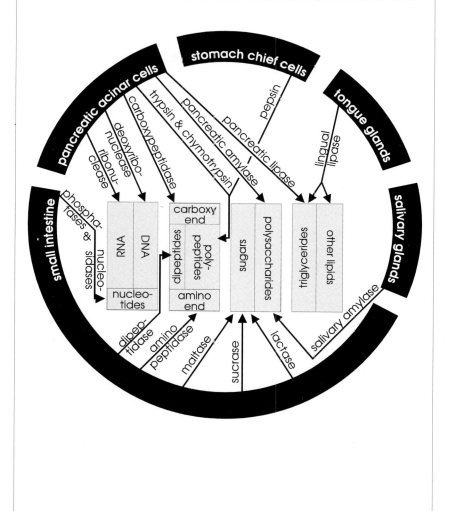

Approximately 90% of nutrient absorption occurs in the small intestine. Most absorbed molecules are transported from capillaries of the villi to the liver by the hepatic portal system. Epithelial cells of the mucosa and the capillary network of the villi absorb monosaccharides by facilitated diffusion. Sugars cotransport with sodium ions, which are actively transported.

Some amino acids and nucleotides are absorbed via cotransport (with Na⁺) across epithelial cell membranes. Other nucleotides and amino acids move into the capillaries by active transport. Some amino acids move by facilitated diffusion.

Most fatty acids and monoglycerides move by simple diffusion into the brush border of the mucosa. Fatty acids may combine with bile salts to form small spheres called **micelles**. Within the epithelial cells, fatty acids are reincorporated into triglycerides. They are packaged in protein-coated aggregates called **chylomicrons**. Chylomicrons move from the villi into the lacteal by exocytosis. They are transported in the lymphatic system, as the mixture, chyle, to the junction of the thoracic duct and the left subclavian vein, where they enter venous circulation.

Water and most vitamins are absorbed by simple diffusion. Minerals are absorbed by passive and active transport mechanisms. Any materials not digested and absorbed by the small intestine pass through the ileocecal valve into the large intestine.

The liver, gallbladder, and pancreas are accessory organs associated with the small intestine. The **liver**, the second largest organ, produces bile and contributes the enzymes of carbohydrate, lipid, and protein catabolism. It is located inferior to the diaphragm on the right side of the abdominal cavity, and consists of 4 lobes. Between the large right and left lobes is the ventral **falciform ligament**, a mesentery. On the inferior border of the falciform ligament, the remnant of the fetal umbilical vein persists as the ligamentum teres, or **round ligament**. On the posterior surface, the inferior vena cava marks the division between the right lobe and the smaller, associated caudate lobe. The quadrate lobe is sandwiched between the associated left lobe and the gallbladder. The **portal hepatis** is the region where the hepatic artery, hepatic portal vein, and hepatic duct enter and exit the liver.

The liver lobes are composed of smaller lobules that consist of hexagonal plates of hepatic cells that radiate out from a central vein. The central veins empty into branches of the hepatic vein, which transports blood to the inferior vena cava. Expanded blood vessels, the **sinusoids**, run throughout the lobules and drain into the central veins. They receive blood from both the hepatic portal vein and the hepatic artery. Within the sinusoids, **Kupffer's cells** phagocytize old blood cells, bacteria, and foreign or toxic materials.

Hepatic cells, or **hepatocytes**, store nutrients, convert them to other compounds, detoxify harmful materials, and produce bile. **Bile** is a yellow-green alkaline solution containing water, bile salts and phospholipids. It emulsifies fats in the small intestine, breaking large fat globules into droplets, increasing their surface area for digestion by **lipases**. From hepatic cells, bile enters small caniculi, or bile capillaries, which feed into bile ducts. At each of the six corners of a liver lobule there is a portal triad, or tract, containing a bile duct and branches of the hepatic artery and hepatic portal vein. Numerous small bile ducts form the right and left hepatic ducts, which join to form the **common hepatic duct**. The common hepatic duct joins the **cystic duct** from the gall bladder to form the **common bile duct**.

The **gallbladder** is a hollow muscular sac 6-8 centimeters long, sitting in a depression between the right and quadrate lobes of the liver. Bile from the liver is stored and concentrated in the gallbladder. Secretion of cholecystokinin by the small intestine triggers contraction of the gallbladder, forcing bile from the cystic duct into the common bile duct. The common bile duct joins the common pancreatic duct in the wall of the duodenum, forming a swelling called the hepatopancreaticduodenal ampulla. The ampulla opens into duodenum via a raised mound, the major duodenal papilla. The **sphincter of Oddi**, a ring of smooth muscle around the common bile duct and pancreatic duct forms a valve that regulates the flow of bile into the duodenum.

The **pancreas** is a loosely aggregated, elongate gland inferior to the stomach. Its head lies in the curve of the duodenum, while its body extends to the left, terminating in a tail. The pancreas has some endocrine function but is primarily an exocrine organ. Groups of cells, **acini**, secrete pancreatic juice. Pancreatic juice flows into small ducts which eventually unite to form the pancreatic duct and a smaller accessory duct that empty into the duodenum.

Pancreatic juice contains bicarbonate ions. The bicarbonate ions buffer the acidic chyme from the stomach and inhibit the action of pepsin.

Pancreatic juice also contains enzymes. These include the carbohydrate-digesting enzyme, **pancreatic amylase**, and several protein-digesting enzymes, **trypsin, chymotrypsin, elastase**, and **carboxypeptidase**. Trypsin is activated by the presence of an intestinal enzyme, enterokinase, and in turn activates the other protein-digesting enzymes. Pancreatic juice also contains **pancreatic lipase** that digests triglycerides and ribo- and deoxyribonucleases that hydrolyze nucleic acids.

Pancreatic secretions are under both neuronal and hormonal control. Parasympathetic fibers of the vagus nerve stimulate pancreatic secretions during the cephalic and gastric phases of digestion. The entrance of acidic chyme into the duodenum results in the release of the hormone secretin from the enteroendocrine cells of the intestine. Secretin, like cholecystokinin, stimulates the production of pancreatic juice.

The **large intestine** consists of the cecum, colon, and rectum. It forms an inverted U about 7 centimeters in diameter and 1.5 meters long that runs from the ileocecal valve of the ileum to the anus. The large intestine is anchored to the dorsal body wall by a mesentery, the mesocolon. The **cecum** extends from the ileocecal valve as a blind pouch terminating in a blind tube-like **vermiform appendix**. An **ascending colon** runs from the cecum and passes up the right side of the body to the underside of the liver. At the hepatic or right colic flexure the colon runs transversely across the abdominal cavity as the **transverse colon**. On the left side of the body, the colon curves downward at the left colic or splenic flexure, continuing as the **descending** and **sigmoid colon**. The terminal part of the large intestine is the **rectum** which ends with the **anal canal** and **anus**. The opening to the anus is ringed with two sphincters, an inner involuntary sphincter composed of smooth muscle, and an external voluntary sphincter containing skeletal muscle. The large intestine resorbs water and electrolytes from undigested materials and eliminates wastes as feces. It does not produce digestive enzymes.

The mucosa of most of the large intestine is relatively smooth, except in the anal canal where it is folded into longitudinal ridges called anal columns. The epithelial layer is simple columnar epithelium designed for water absorption. Mucus secreting goblet cells produce lubricating mucus to facilitate the movement of the consolidated chyme. The muscularis layer is modified with longitudinal fibers in three muscular bands called the **taeniae coli**. As these bands contract, the large intestine is pulled into numerous pouches called **haustra**. The epiploic appendages are small fat-lined pouches of visceral peritoneum attached to the taenia coli.

Bacteria, or **intestinal flora**, residing in the lumen of the large intestine break down carbohydrates and proteins in the chyme. These materials are absorbed through the mucosa and transported to the liver. Intestinal flora also produce essential vitamins such as B and K. Byproducts of bacterial metabolism include various gasses which are eliminated as **flatus** or gas, and simple molecules such as indole and skatole which are eliminated in the feces.

Chyme moves as distension of the large intestine and contraction of the muscularis result in haustral churning. Chyme also moves through the large intestine by peristalsis, although the rate of peristalsis here is slower than elsewhere in the GI tract. Food entering the stomach stimulates the gastrocolic reflex, and any chyme that has reached the transverse colon is rapidly moved into the rectum by a wave of mass peristalsis. At this point, the material, termed feces, consists of indigestible materials, end products of bacterial metabolism, and dead bacteria.

The presence of feces in the rectum causes a defecation reflex as stretch receptors in the walls of the rectum are stimulated. Parasympathetic nerve stimulation of rectal wall muscles and relaxation of the internal sphincter create a feeling of urgency. Voluntary relaxation of the external sphincter, in concert with contraction of the diaphragm and abdominal muscles, result in defecation, or expulsion of feces from the body via the anus.

Chapter 9

THE REPRODUCTIVE SYSTEM

The reproductive system (1) produces haploid gametes, eggs or sperm, (2) stores, transports, and sustains the gametes, (3) provides for internal fertilization, (4) nurtures the developing fetus in females, and (5) produces hormones that are responsible for secondary sexual characteristics, development of the reproductive system, and gamete maturation. The reproductive system consists of gonads which are the sites of gamete and hormone production, tubes for transport of gametes and reproductive products, and accessory glands.

The **testes**, the male gonads, are the sites of sperm production. These paired oval glands are about 5 centimeters long and 2.5 centimeters in diameter. They arise from the same embryonic tissue as ovaries, and begin their development in the abdominal cavity, near the kidneys. During the seventh month of fetal development, each testis descends through an opening in the inferior body wall, the **inguinal canal**, into the scrotum. The **scrotum**, part of the external genitalia, is a skin-covered pouch that houses the testes and the epididymis. Scrotal temperature is 3 degrees centigrade cooler than core body temperature, a condition critical for normal development and survival of sperm. The scrotum is divided into two scrotal cavities by the **scrotal septum**, a vertical sheet of superficial fascia and dartos muscle fibers. A testis resides in each cavity. The scrotal sac has several layers. Internal to the skin is a layer of **fascia dartos** muscle that wrinkles the scrotum. The cremasteric layer is a band of skeletal muscle. Contraction of **cremasteric fibers** regulate scrotal temperature by drawing the scrotal sac closer to the body in cold weather. The innermost **tunica vaginalis** consists of serous membrane and is continuous with the parietal peritoneum of the abdominal cavity.

SAGITTAL SECTION OF MALE REPRODUCTIVE ORGANS

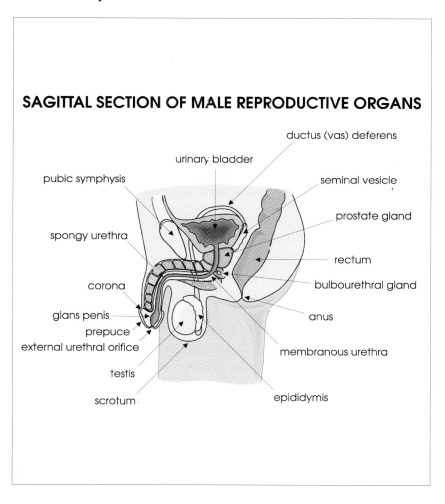

A white fibrous connective tissue layer, the **tunica albuginea**, covers the surface of each testis. This layer infolds into numerous septa that divide each testis into about 250 small lobules. Within each lobule are 1 to 4 highly coiled seminiferous tubules, the sites of sperm production. Hormone-secreting interstitial endocrinocytes, or **Leydig cells**, occur in spaces between tubules.

Seminiferous tubules from each lobule feed into a straight duct, the **tubulus rectus**, that conveys sperm into the **rete testes**, a network of tubules on the posterior of the testis. From the rete testes, efferent ducts transport sperm to the epididymis. The **epididymis** are paired structures on the posterior border of each testis. Each epididymis consists of the **ductus epididymis**, a highly coiled tube about 7 meters in length. The tube has three segments; superior head, body and tail, which connects with the ductus deferens. The epididymis provides for final sperm maturation and storage, monitors and adjusts the composition of seminal fluids, and destroys aged sperm. During ejaculation, peristaltic contractions of the smooth muscle fibers lining the walls of the ductus epididymis propel sperm and a small amount of seminal fluid into the ductus deferens. Sperm and accessory fluids secreted by the reproductive tract comprise the **semen**.

The **ductus deferens** or **vas deferens** is a fibromuscular tube about 40 centimeters in length. Each duct carries spermatozoa from the epididymis to an ejaculatory duct during ejaculation. The ductus deferens runs from the posterior aspect of the epididymis superiorly through the inguinal canal and into the body cavity, passing over a ureter and descending along the dorsal surface of the urinary bladder. At its terminus, the ductus is enlarged to form the ampulla, and is joined by a duct from the seminal vesicle. The duct continues as a short tubule, the **ejaculatory duct**, which enters the single urethra at the level of the prostate gland. The **spermatic cord** consists of the ductus deferens, a testicular artery and vein, autonomic nerve fibers, lymph vessels, and cremasteric muscles running from the scrotum through the inguinal canal.

The **urethra** is a single tube connecting the urinary bladder to the tip of the penis. The prostatic urethra passes through the prostate gland, the membranous urethra passes through a muscular sheet, the **urogenital diaphragm**, and the spongy urethra passes through the penis, opening to the outside at the external urethral orifice. In males, the urethra transports urine and sperm. A sphincter at the base of the bladder constricts to prevent the release of urine during ejaculation.

The urethra is surrounded by the intromittent organ or **penis**. The body of the penis consists primarily of three longitudinal columns of spongy vascular tissues that cause the penis to become erect when engorged with blood. The **corpus spongiosum** of the penis surrounds the spongy urethra. Two **corpora cavernosa** are located dorsal and lateral to the corpus spongiosum. The distal end of the corpus spongiosum expands as the **glans penis**, the location of the external urethral orifice. The highly sensitive glans is covered by a cuff of skin, the **prepuce** or foreskin, which is surgically removed in circumcision.

The penis root attaches to the pelvis. The **crura**, the expanded proximal ends of the corpora cavernosa, attach to the ischial and pubic rami. They are associated with the **ischiocavernosus muscle**, which contracts to force blood from the crura into the corpora cavernosa during erection. The bulb of the penis is the expanded base of the corpus spongiosum. It is attached to the urogenital diaphragm and surrounded by the **bulbospongiosus muscle**. Its contraction compresses the bulb, emptying the urethra of residual urine or semen.

Accessory glands produce and secrete the liquid component of semen. Paired seminal vesicles, dorsal to the urinary bladder base, release fluid into the ejaculatory ducts. This alkaline fluid containing fructose, prostaglandins and fibrinogen comprises 60% of the liquid component of semen. It buffers acidic conditions in the female reproductive tract and provides an energy source for sperm locomotion.

The single ring-shaped **prostate gland** surrounds the prostatic urethra at the base of the bladder. The prostate opens into the urethra by a series of small ducts. Its milky secretion is slightly acidic, containing citric acid, and comprises about 1/3 of the seminal fluid. It also contains enzymes, antibacterials, and substances enhancing sperm activity and survival.

The **bulbourethral glands** are paired pea-sized structures within the urogenital diaphragm, with ducts opening into the spongy urethra. Their alkaline mucus secretion, released prior to ejaculation, lubricates the penis and cleans the urethra of traces of urine.

The female reproductive system provides an environment for development of the fertilized egg and a pathway for the fully-developed fetus to exit its mother. The female gonads, the **ovaries**, produce ova (eggs) and sex hormones. These paired ovate structures are roughly 3.5 centimeters in length and about 2 centimeters wide, situated in the upper pelvic cavity in shallow depressions called **ovarian fossae** on either side of the uterus. Each ovary has a hilus, a recessed area where the ovarian artery and vein enter and exit. Ovaries are anchored by mesenteries. Two mesenteries, the **suspensory ligament** and **mesovarium**, are continuous with peritoneum attached to the uterus by the broad ligament. Ovarian blood vessels course through the **suspensory ligament**. A fibrous band within the broad ligament, the **ovarian ligament**, runs from the medial aspect of each ovary to the uterus. The external surfaces of the ovaries are covered by simple cuboidal epithelium, the germinal epithelium or visceral peritoneum, continuous with the mesovarium. The fibrous **tunica albuginea** lies beneath this layer. Deep to it, the stroma is divided into the cortex, where ovarian follicles and ova are located, and an inner medulla of loose connective tissue.

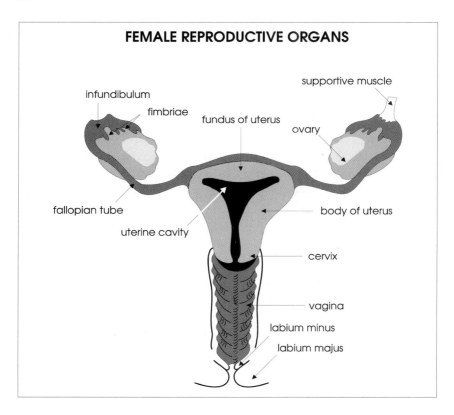

FEMALE REPRODUCTIVE ORGANS

The **oviducts**, or **uterine** or **Fallopian tubes**, are paired tubes about 12 centimeters long, connecting the ovaries to the uterus. The **infundibulum**, the wide funnel-like opening near each ovary is ringed by finger-like fimbriae. **Fimbriae** sweep over the ovary's surface and carry ovulated eggs into the oviduct by producing a current in the peritoneal fluid. The main portion of the oviduct is the ampulla, the site of fertilization, which constricts to form the **isthmus** shortly before joining the uterus. The inner mucosa layer of the oviduct is ciliated epithelium. Movement of the cilia and contraction of smooth muscle in the oviduct walls move the egg through the tube, a journey of 3 to 4 days.

The **uterus** is a roughly triangular muscular sac posterior to the bladder and anterior to the rectum. The **fundus** of the uterus is the superior region where the oviducts enter. The **body** is the main portion of the uterus while the **cervix**, or neck, is the constricted base, tilting posteriorly. Spaces inside these regions are the uterine cavity and cervical canal, respectively. Because of the posterior angle of the cervix, the body of the uterus tilts anteriorly over the urinary bladder (**anteflexion**). It is secured by several band-like mesenteries called ligaments. Broad ligaments form a sheet over the uterus and extend to the lateral pelvic wall. Round ligaments are thick bands embedded within the broad ligaments, and contain smooth muscle, blood vessels and nerves. Uterosacral ligaments connect the uterus to the sacrum.

The uterine wall has three layers. The innermost endometrium is composed of two sublayers. The inner **stratum functionalis** is a vascular tissue that forms the maternal contribution to the placenta upon fertilization. It arises from a growth layer, the **stratum basale**, and is lost monthly as menstrual flow if fertilization does not occur. Uterine arteries, which branch into arcuate and then radial arteries before terminating in spiral arterioles, supply blood to the stratum functionalis. The central layer, or **myometrium**, consists of three smooth muscle sublayers. The myometrium undergoes strong rhythmic contractions during labor. The outermost layer, the serosa or perimetrium, is continuous with the broad ligament.

Between the bladder and rectum is the **vagina**, an elastic muscular tube about 10 centimeters long, extending from the cervix to the exterior vaginal orifice. The vaginal orifice is partially covered by a thin mucous membrane, the **hymen**. At its superior end, the vagina forms a pouch-like fornix around the cervical lip. The inner lining, the mucosa, is stratified squamous epithelium, and produces acidic secretions that retard bacterial growth. The muscularis layer contains smooth muscle fibers. The vagina expands to accommodate the erect penis during coitus, and to permit passage of the baby during childbirth. The external vaginal orifice and urethral orifice are surrounded by external genitalia, the **vulva** or pudendum. An anterior, hair-covered pad of adipose tissue, the **mons pubis**, cushions the pubic symphysis. Running posteriorly from the mons to the anus, lateral to the urethral and vaginal orifices, are two pairs of folds, the labia majora and minora, containing sebaceous and sweat glands. The outer **labia majora**, homologous to the scrotum, are two fatty, hair-covered folds of skin. The medial folds, the **labia minora**, are delicate and hairless. The region between them is the **vaginal vestibule**. The labia minora are homologous to the shaft of the penis and contain erectile tissue. They converge anteriorly to form a fold of skin, the prepuce or foreskin, which covers the **clitoris**, homologous to the glans penis. Stimulation of the clitoris can result in sexual arousal and orgasm, contraction of smooth muscle of the reproductive tract.

Accessory glands include paraurethral or Skene's glands, and greater vestibular or Bartholin's glands. The **paraurethral glands** are homologous to the prostate gland, and secrete lubricating mucus from the base of the urethra into the vaginal vestibule. The **greater vestibular glands**, homologues of bulbourethral glands, secrete into the vaginal vestibule at the level of the hymen.

Spermatogenesis is production of sperm cells, or spermatozoa, in the seminiferous tubules of the testes. The tubules contain two cell types, **sustentacular cells**, which aid sperm development, and spermatogenic cells, which give rise to sperm. **Spermatogonia** are diploid cells, each with 46 chromosomes, located on the periphery of the seminiferous tubules. At puberty, hormones stimulate spermatogonia to multiply by mitosis. Some of the daughter cells continue to divide mitotically providing a constant supply of spermatogonia. Other daughter cells become primary spermatocytes which undergo meiosis I, producing two secondary spermatocytes. Each secondary spermatocyte undergoes meiosis II, producing two haploid spermatids with 23 chromosomes each. **Spermiogenesis** is the process in which spermatids attach to sustentacular cells and mature into spermatozoa. A spermatozoan has a headpiece containing chromosomes and an acrosomal cap filled with enzymes necessary for fertilization. The tail or flagellum provides for locomotion. The flagellum and head are connected by the midpiece. Mitochondria in the midpiece produce ATP necessary for locomotion.

The large **sustentacular**, or Sertoli, cells extend from the walls of the tubules into the lumen. They support cell reproduction, aid spermiogenesis, and secrete the hormones inhibin and androgen-binding protein. Sustentacular cells maintain the blood/testis barrier, preventing destruction by the immune system of developing sperm. Fully developed spermatozoa are released by sustentacular cells, and move into the lumen of the tubules, a process called **spermiation**. They pass to the epididymis, where after two weeks they have become fully mature and motile. The entire process takes approximately 10 weeks.

Five hormones control male reproduction. **Gonadotropin releasing hormone** (GnRH) is secreted by the hypothalamus into the hypothalamic-pituitary portal vessels and is carried to the anterior pituitary gland. GnRH stimulates the release of two gondatropins, **follicle stimulating hormone** (FSH) and **luteinizing hormone** (LH), which are carried in the blood to target cells in the testes. Under the influence of LH and FSH, Leydig cells secrete the male sex hormone, **testosterone**. Testosterone is responsible for male sex drive and, in high levels, inhibits release of GnRH and LH. FSH and testosterone also trigger responses from sustentacular cells that stimulate spermatogenesis. Secretion of **inhibin** by sustentacular cells exerts negative feedback, depressing the release of FSH.

Male secondary sex characteristics develop at puberty under the influence of androgens, but are not directly involved in sexual reproduction. Examples include muscular and skeletal development resulting in wide shoulders and narrow hips, growth of pubic, underarm, facial, and chest hair, and enlargement of the larynx resulting in deepening of the voice.

In females, production of ova by ovaries, or **oogenesis**, begins during early fetal development, as diploid oogonia undergo mitosis and differentiate into primary oocytes. Primary oocytes enter meiosis I, but stop at prophase I and remain in that state until puberty. At birth, each ovary contains several hundred thousand primary oocytes, each surrounded by a single layer of ovary cells, constituting the primordial follicle. At puberty, the ovaries become active in response to FSH and LH resulting in the **ovarian cycle** of egg maturation and release, and the uterine or **menstrual cycle** of uterine preparation for pregnancy. The onset of these monthly cycles is the first period of menstrual bleeding, or **menarche**. **Menopause**, or cessation of the reproductive cycles usually occurs when a woman is in her 40s or 50s.

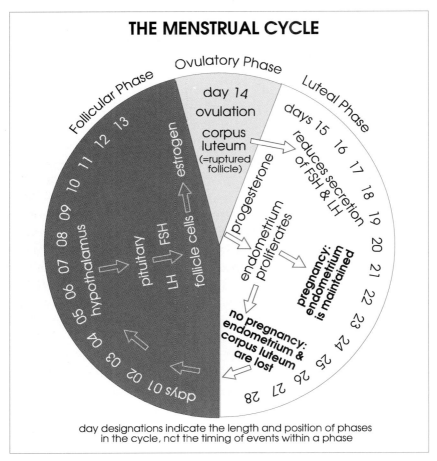

THE MENSTRUAL CYCLE

day designations indicate the length and position of phases in the cycle, nct the timing of events within a phase

The **ovarian cycle** begins with the preovulatory or follicular phase. Under the influence of GnRH, FSH and LH are secreted by the pituitary gland and stimulate development of about 25 primary follicles, each a primary oocyte surrounded by several layers of follicular cells called **granulosa cells**. With further development, many of the primary follicles enlarge to become secondary follicles. A thick layer of glycoproteins, the **zona pellucida**, develops around the ovum. As the granulosa cells enlarge and multiply, cells in the ovarian stroma differentiate into a layer of thecal cells surrounding the follicle. These layers secrete the female sex hormones, the estrogens, and inhibin. **Estrogen** stimulates the development of female sex organs and stimulates new growth cf the stratum functionalis of the uterus. Follicle cells secrete follicular fluid into the cavity, or **antrum**, between the ovum and follicle. Usually only one primary follicle will complete the cycle while others undergo **atresia**, or breakdown. The remaining follicle develops into a mature **Graafian**, or vesicular ovarian **follicle** while the primary oocyte completes meiosis I, producing a secondary oocyte and a non-functional polar body. The secondary oocyte enters meiosis II, but is arrested in metaphase.

The second phase of the ovulatory cycle, **ovulation**, occurs about 14 days into the cycle as the follicle ruptures in response to a surge of LH. The secondary oocyte and a ring of granulosa cells, the **corona radiata**, pass from the ovary into the oviduct. The post-ovulatory phase lasts approximately 14 days, from ovulation until menstruation. The remaining granulosa cells of the ruptured follicle, the **corpus luteum**, continue to secrete estrogens but also begin to secrete progesterone. **Progesterone** accelerates proliferation of the endometrium in preparation for implantation of a fertilized egg. With estrogen, progesterone inhibits release of GnRH, LH, and FSH. If fertilization does not occur, the corpus luteum degenerates to a white scar tissue mass, the **corpus albicans**. Levels of progesterone and estrogen fall, allowing hypothalamic and pituitary hormone levels to increase again, beginning a new ovarian cycle.

The uterine cycle entails monthly changes in the endometrium. The **menstrual phase**, coinciding with the follicular phase, is a 5 day period in which the stratum functionalis and about 50 milliliters of blood are lost as menstrual flow through the vagina. This occurs as disintegration of the corpus luteum causes progesterone and estrogen levels to fall. In the proliferative phase, lasting about 8 days, secretion of estrogen by the granulosa cells of the ovarian follicle stimulates the stratum basale to begin production of a new stratum functionalis layer.

In the secretory phase, secretion of progesterone by the corpus luteum results in thickening of the endometrium, which secretes glycogen and other substances. If pregnancy occurs, the endometrium is maintained, comprising the maternal contribution to the placenta. If no egg is fertilized, the cycle begins anew as the stratum functionalis breaks free and is lost as menstrual flow.

Coitus, or sexual intercourse, is the mechanism by which fertilization occurs. Parasympathetic stimulation results in penile erection, caused by dilation of the arterioles feeding the vascular columns of the penis, and the engorgement of those tissues with blood. The bulbospongiosus and ischiocavernosus muscles contract, compressing the penile veins and impeding the drainage of venous blood. Sufficient stimulation of the glans penis in particular will result in ejaculation, or orgasm, in two phases. In emission, sympathetic stimulation results in contraction of smooth muscle fibers lining all reproductive structures exclusive of the urethra. Expulsion of the semen out of the urethra is caused by spasmodic contractions of the musculature of the penis. The volume of ejaculate may range from 2-6 milliliters, and contains 125 to 750 million sperm.

Sexual arousal in females results in mucus secretions of the vestibular glands and cervical epithelium, lubricating the vagina for penile penetration. Female orgasm may occur as a result of coitus, but is not necessary for fertilization. Semen enzymes cause fibrinogen in the semen to clot once it is deposited within the female vagina. Further enzymatic action causes the clot to liquefy within 20 minutes. Sperm travel rapidly from the vagina through the uterus and into the oviducts. If an ovum is present in the ampulla, fertilization may occur.

As sperm contact the egg, enzymes in the acrosome begin digesting the corona radiata and zona pellucida. Although only one sperm fertilizes an ovum, many are necessary to dissolve the zona pellucida. Once a sperm penetrates the egg cell membrane, leaving its flagellum on the exterior, the ovum completes meiosis II, forming a haploid nucleus and a non-functional polar body. Penetration of the ovum by a sperm results in chemical changes that prevent other sperm from entering. Fusion of ovum and sperm nuclei, or fertilization, produces a single diploid cell, the **zygote**.

As the zygote moves down the oviduct, the first mitotic division, or cleavage, occurs about 36 hours after fertilization. Cleavage continues at 12 hour intervals, forming a solid ball of cells, the **morula**, that moves into the uterus. By the fifth day after fertilization, the zona pellucida breaks down and cells rearrange into the **blastocyst**, a hollow sphere with an inner cell mass. Implantation occurs as the outer ring of cells, the **trophoblast**, burrows into the uterine endometrium. The trophoblast becomes the **chorion**, a fetal membrane that with the decidua, or endometrium, forms the **placenta**. The placenta permits nutrient and gas exchange between developing embryo and mother. The trophoblast begins secreting the hormone, **human chorionic gonadotropin**, that initially maintains the corpus luteum, insuring elevated progesterone and estrogen levels. By the eighth week of pregnancy, the placenta secretes these hormones and the role of the corpus luteum declines.

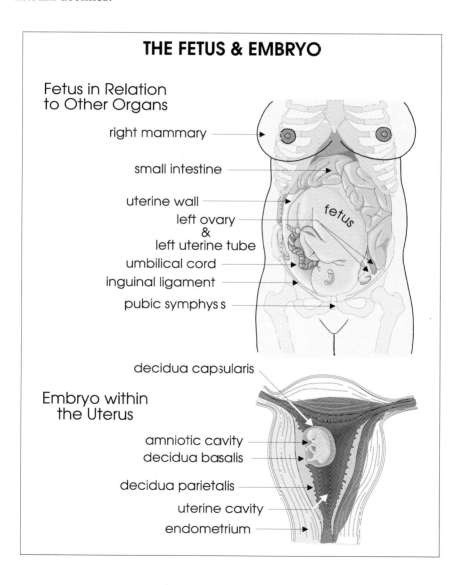

THE FETUS & EMBRYO

Fetus in Relation to Other Organs

- right mammary
- small intestine
- uterine wall
- left ovary & left uterine tube
- umbilical cord
- inguinal ligament
- pubic symphys s

fetus

Embryo within the Uterus

- decidua capsularis
- amniotic cavity
- decidua basalis
- decidua parietalis
- uterine cavity
- endometrium

Several other extraembryonic membranes are associated with the embryo. The **amnion** is a thin-walled, fluid-filled sac enclosing the embryo by the tenth day of pregnancy. The **amniotic fluid** within the amniotic cavity cushions and protects the embryo. The **yolk sac** outpockets from the gut region. It is the source of primordial germ cells which migrate into the gonads. The **allantois**, an extension off the yolk sac, is a site of blood formation. It later contributes to the **umbilical cord**, connecting the placenta to the embryo.

By the end of the first three months (the **first trimester**), the placenta is established and all major organs and limb buds are forming. Now, the embryo is termed a **fetus**. By the end of the **second trimester**, development of organ systems is well underway, and skeletal muscle development has begun. At the end of the **third trimester**, when organ systems are complete, hormones are secreted, particularly **oxytocin** from the pituitary gland, that intiate wave-like contractions of the uterus, or labor pains. The hormone **relaxin** relaxes fibers in the pubic symphysis, and the cervix enlarges or dilates. Rupturing of the amnion also occurs at this time. In the final stages of labor, the fetus is pushed through the vagina or birth canal, and is expelled from the body in the expulsion phase of labor. Soon after, the placenta or afterbirth is expelled in the placental phase of labor.

The baby may be nursed by the mother's **mammary glands**. These modified milk-producing sudoriferous glands form fatty pads external to the pectoralis muscles to which they are attached by connective tissue. A pigmented **areola** surrounds the nipples which contain both tactile sensors and smooth muscle. Nipples become erect when stimulated.

Internally, the adult female breast has lobes made of smaller lobules. Lobes are separated from one another by fat tissue. Lobules are clusters or alveoli of milk-secreting glands and connective tissue. With **lactation**, milk flows from the glands via **lactiferous ducts** to the nipples. Within the **nipples**, the ducts expand into lactiferous sinuses.

Development of mammary glands begins at puberty with the secretion of female sex hormones. Several hormones control milk production, or lactation, during pregnancy. Estrogen increases breast size by enhancing fat deposition. Progesterone causes proliferation of the duct system. **Prolactin** stimulates the glands to produce milk after the baby is born. The pituitary secretes prolactin before birth, but estrogen and progesterone inhibit milk production then. The sucking action of the infant triggers the hypothalamus to stimulate release of oxytocin from the posterior pituitary gland, resulting in the flow of milk to the nipples.

Index

A

acinar cells 9
acini 95
adenoids 52, 62
adipose capsule 72
adrenal cortex 6, 7
adrenal glands 6
adrenal medulla 6, 9
adrenaline 9
adrenocortical hormones 2
adrenocorticotrophic hormone 4, 7
adrenocorticotrophic releasing hormone 7
adventitia 82, 83
afterbirth 110
agglutinins 17
agranulocytes 19
albumins 13, 14
aldosterone 7, 80, 81
alimentary canal 84
allantois 110
alpha cells 9
alveolar pressure 65
alveoli 63
amnion 110
amniotic fluid 110
anal canal 96
androgens 8, 11
angina pectoris 27
angiotensin converting enzyme 7
angiotensin II 7, 37, 81

angiotensinogen 7
anteflexion 103
anterior pituitary gland cells 4
antibodies 59
antidiuretic hormone 4, 37, 80, 81
antigen-presenting cells 57
antimicrobial chemicals 53
antrum 107
anus 96
aorta
    abdominal 41
    thoracic 40
aortic arch 39
arteries 33
    anterior tibial 43
    arcuate 76
    basilar 40
    brachiocephalic 39
    celiac trunk 41
    common hepatic 41
    common iliac 43
    conducting 33
    deep femoral 43
    dorsalis pedis 43
    external iliac 43
    femoral 43
    gastric 41
    gastroduodenal 41
    gonadals 42
    hepatic 41
    inferior mesenteric 42
    interlobar 76
    interlobular 76
    internal carotids 39
    internal iliac 43
    left common carotid 39
    left gastric 41
    left gastroepiploic 41
    lumbar 42
    pancreatic 41
    peroneal 43
    plantar 43
    popliteal 43
    posterior tibial 43
    pulmonary 31
    renals 42
    right gastric 41
    splenic 41
    subclavian 39
    superior mesenteric 42
    suprarenal 42
    systemic 31

umbilical 48
vertebral 40
arterioles 32, 33
afferent 75, 77
efferent 75, 77
asthma 63
atresia 107
atria 23
atrial natriuretic hormone 11, 81
atrial natriuretic peptide 37
atriopeptin 11, 81
atrioventricular valves 25
bicuspid 25
tricuspid 25
auricle 23
auscultations 29
azygos system 46

B

B-cell activation 59
B-cells 51, 52, 56, 57, 59
memory 59
basophils 19
beta cells 9
bicarbonate ion 14, 70, 80, 95
bile 95
bilirubin 16
birth canal 110
blastocyst 108
blood 12
blood clotting 20
blood colloid osmotic pressure 35
blood hydrostatic pressure 34
blood pH 80
blood pressure 34, 35, 37, 38
normal 38
blood reservoirs 35
blood type 17
body of the uterus 103
bolus 88
Bowman's capsule 74
Boyle's law 65
bradycardia 38
breathing 65
broad ligaments 103
bronchi
lobar 63
primary 63, 64
secondary 63
bronchioles 63
bronchopulmonary segments 63
buffy coat 19

bulbospongiosus muscle 101
bulbourethral glands 101
bundle of His 27

C

calcitonin 6
calcitriol 81
calyces
major 73
minor 73
canines 87
capillaries 32, 33
peritubular 77
capillary exchange 34
carbaminohemoglobin 16
carbon dioxide 14, 16, 31, 67, 70
carbonic acid 70, 80
carbonic anhydrase 14, 80
carboxypeptidase 96
cardiac cycle 28
cardiac output 29
caries 88
carina 63
cartilage
arytenoid 62
corniculate 63
cricoid 62
cuneiform 63
thyroid 62
catecholamines 2
cecum 96
cell-mediated immunity 57
cementum 88
cervix 103
chemotaxis 54
chief cells 89
cholecystokinin 11, 90, 92
cholesterol 2
chorion 108
chromaffin cells 9
chyle 51
chylomicrons 94
chyme 89, 90, 91, 95, 97
chymotrypsin 96
circulation
fetal 38, 48
pulmonary 31, 38
systemic 31, 38
circumcision 101
clitoris 104
clotting factors 20, 21, 54
coagulation 20

coagulation factors   13
coitus   108
collecting duct   76
colon   96
  ascending   96
  descending   96
  sigmoid   96
  transverse   96
common bile duct   95
common hepatic duct   95
complement system   53
conchae   61
corona radiata   107, 108
coronary arteries   25
coronary sinus   23, 26
coronary sulcus   23
corpora cavernosa   101
corpus albicans   107
corpus luteum   11, 107, 109
corpus spongiosum   101
cortex of the kidney   73
corticotrophs   4
corticotropin releasing hormone   7
cortisol   7
counter current mechanism   80
creatinine   78, 82
cremasteric fibers   99
cricothyroid ligament   62
crown   88
crura   101
crypts of Lieberkühn   91
cycle
  menstrual   105
  ovarian   105, 106, 107
  uterine   105, 107
cycterna chyli   51
cystic duct   95
cytokines   59

D

Dalton's law   68
decidua   108
defecation   97
defense mechanisms
  nonspecific   53
deglutition   88
delta cells   9
dendritic cells   57
dentin   88
dentition   87
detrusor muscle layer   83
diabetes mellitus   80

diapedesis   54
diaphragm   65
diastole   28
digestion   84
digestive system   84
distal convoluted tubule   75
ductus arteriosus   48
ductus deferens   100
ductus epididymis   100
ductus venosus   48
duodenum   90

E

edema   35
eicosanoids   2
ejaculatory duct   100
elastase   96
electrocardiogram   28
embolus   21
embryo   110
emphysema   68
enamel   88
endocardium   23
endocrine glands   1
endocrine system   1
endometrium   107, 108
enteroendocrine cells   89, 91
eosinophils   19, 54
epicardium   22, 23
epididymis   100
epiglottis   62
epinephrine   2, 9, 37
erythroblastosis fetalis   18
erythrocytes   14
erythropoiesis   16
erythropoiesis stimulating hormone   16
erythropoietin   16, 37, 75, 81
esophagus   89
estrogen   2, 107, 110
eustachian tube   62
exocrine glands   1
expiration   65, 67
external intercostal muscles   65

F

F cells   9
falciform ligament   94
Fallopian tubes   103
fascia dartos   99
fauces   62, 86
feces   85, 97
ferritin   16
fetal hemoglobin   69

fetus 110
fever 54
fibrin 20
fibrinogen 14, 20
fibrinolysis 20
fimbriae 103
flatus 97
follicle stimulating hormone 4, 9, 11, 105
foramen ovale 23, 48, 49
formed elements of blood 12, 14
fossa ovalis 23, 49
fundus 89, 103

G

gallbladder 95
gametes 98
gastric glands 89
gastric inhibitory peptides 11, 90
gastric pits 89
gastrin 90
GI tract 84, 86
gingivae 86
glans penis 101, 104
globin 15, 16
globulins 13
  alpha 14
  beta 14
  gamma 14, 59
glomerular capsule 74
glomerulus 74
glottis 62
glucagon 9
glucocorticoids 7
goblet cells 63
goiter 6
gonadocorticoids 8
gonadotrophins 4
gonadotrophs 4
gonadotropin releasing hormone 9, 105
gonads 9, 98
Graafian follicle 107
granulocytes 19, 54
granulosa cells 106
greater vestibular glands 104
growth hormone-inhibiting hormone 9
growth inhibiting hormone 4
growth releasing hormone 4

H

haustra 97
heart 22, 23, 25, 26, 27, 28, 29, 30
heart contractions 27
heart murmurs 29

heart rate 29, 30
hematocrit 16
hematopoiesis 16
heme 15
hemocytoblasts 16
hemoglobin 14, 15, 16, 69
hemophilia 21
hemopoiesis 16
hemosiderin 16
hemostasis 20
Henry's law 68
heparin 16
hepatic portal system 47
hepatocytes 95
hepatopancreaticduodenal ampulla 95
hilus 51, 64, 73
histamine 54
homeostasis 71
human chorionic gonadotropin 109
human chorionic gonadotropin hormone 11
human growth hormone 4
humoral immunity 59
hydrocortisone 7
hymen 103
hypophysis 2
hypoxia 16, 27

I

ileocecal valve 91
ileum 91
immune response 59
immune system 55
immunoglobulins 14, 59
immunological surveillance 54
incisors 87
incontinence 83
infarct 21
inferior vena cava 23, 43, 46, 77
inflammation 54
infundibulum 2, 103
inguinal canal 98
inhibin 11, 105
inner zona reticularis 8
inspiration 65
inspiratory capacity 68
insulin 9
interatrial septum 23
intercalated cells 80
interferons 53
intestinal flora 97
intracranial vascular sinuses 43
intromittent organ 101

iron   16, 69
ischiocavernosus muscle   101
islets of Langerhans   9
isthmus   103

J

jejunum   91
juxtaglomerular apparatus   75
juxtaglomerular cells   7, 75, 81

K

keratin   53
ketone bodies   82
kidneys   71, 72
Kupffer's cells   95

L

labia
    majora   104
    minora   104
labial frenulum   86
lactation   110
lacteals   50
lactotrophs   4
lamina propria   82, 85
large intestine   96
laryngopharynx   62, 88
larynx   62
left atrium   23
leukocytes   19
leukotrenes   2, 54
Leydig cells   100, 105
ligamentum arteriosum   49
ligamentum venosum   49
lingual frenulum   87
lipases   95
lipolysis   7
liver   94
lobule   63
loop of Henle   75, 77, 78, 79, 80
lumen   32
lungs   64, 65
luteinizing hormone   4, 9, 11, 105
lymph   35, 50, 51
lymph nodes   51
lymphatic organs   51
lymphatic system   50
lymphoblasts   19
lymphocytes   19, 51, 52, 56
lymphoid stem cells   51
lysozymes   53

M

macrophages   51, 52, 56, 57
    fixed   54
    wandering   54
macula densa   75
major histocompatibility complex antigens
57, 59
mammary glands   110
mast cells   2, 54
medial umbilical ligaments   48
mediastinum   64
megakaryoblasts   19
melanocyte-stimulating hormone   4
melatonin   2
menarche   106
menopause   106
mesovarium   102
metabolic wastes   78
micelles   94
microphages   54
microvilli   78, 91
micturition reflex   83
middle zona fasciculata   7
mineralocorticoids   7
molars   87
monoblasts   19
monocytes   19, 54
mons pubis   104
morula   108
mucosa   82, 85
mucus   53
mucus-secreting cells   89
muscularis   82
muscularis externa   86
muscularis mucosa   86
myeloblasts   19
myocardial infarction   21, 27
myocardial ischemia   27
myocardium   23
myometrium   103

N

nasal cavity   60
nasal septum   60
nasal turbinates   60
nasopharynx   62, 88
natural killer cells   51, 54
navel   48
nephron   74
    cortical   77
    juxtamedullary   77
neutrophils   19, 54
nipples   110

noradrenaline 9
norepinephrine 2, 9, 37
nose 60

O

oocyte 105, 106, 107
oogenesis 105
oral cavity 86
oropharynx 62, 87, 88
ova 9, 101, 105
ovarian fossae 102
ovarian ligament 102
ovaries 9, 101, 102, 105
oviducts 103
ovulation 107
oxygen 14, 15, 16, 31, 67, 69, 70
oxyhemoglobin 14, 15, 69
oxytocin 4, 110, 111

P

palate
    hard 86
    soft 86, 87
pancreas 9, 95
pancreatic amylase 96
pancreatic juice 95
pancreatic lipase 96
pancreatic polypeptide 9
papillary ducts 76
parafollicular cells 6
paranasal sinuses 61
parathormone 6
parathyroid glands 6
parathyroid hormone 6
paraurethral glands 104
parietal cells 89
parotid gland 88
penis 100, 101, 108
pepsin 95
peptide hormones 1, 2
perforin 54
pericardium 22
peristalsis 82, 86, 97
peritonitis 86
Peyer's patches 52, 91
phagocytes 54
phagocytosis 53
pharynx 62, 88
physical barriers 53
pituitary gland 2
placenta 11, 48, 107, 108, 110
plasma 12, 17, 71
plasma cells 52, 59

plasmin 20
plasminogen 20
platelets 19
pleura 65
pleural membrane 65
pleurisy 65
pleuritis 65
plexus of Auerbach 86
plexus of Meissner 86
plicae circulares 91
podocytes 75
portal hepatis 94
potassium 80, 81
premolars 87
prepuce 101
principal cells 80
proerythroblasts 16
proerythrocytes 16
progesterone 11, 107, 110
prolactin 4, 110
prostaglandins 2, 54
prostate gland 100, 101
proteins
    CD4+ 56
    CD3+ 56
prothrombin 20, 21
proximal convoluted tubule 75
pulmonary ventilation 65
pulse 37
pulse pressure 38
pulse rate 37
    resting 38
Purkinje fibers 27
pus 54
pyloric antrum 89
pyloric canal 89
pyloric region 89
pyloric sphincter 89

R

rectum 96, 97
red blood cells 14, 16, 52
    immature 16
red bone marrow 51
regulating-hypothalamic hormones 2
relaxin 110
renal capsule 73
renal columns 73
renal corpuscle 74
renal fascia 72
renal papilla 73, 76
renal pelvis 73

renal pyramid 73
renal sinus 73
renal tubule 75
renin 7, 37, 75, 81
reproductive system 98
reserve volume
    expiratory 68
    inspiratory 68
residual volume 68
resistance 52
respiration
    external 68
    internal 68
respiratory membrane 63
respiratory system 60
    upper 60
resting breathing rate 68
rete testes 100
retention 83
reticulocytes 16
Rh blood group 17
right atrium 23
right lymphatic duct 51
rima glottidis 62
roots 88
round ligament 94, 103
rugae 82, 89

S

salivary amylase 88
salivary glands 88
scrotal septum 99
scrotum 98, 99
sebum 53
secretin 11, 90, 92, 96
sella turcica 2
semen 100, 108
semilunar valves 25
    aortic 25
    pulmonary 25
seminal vesicles 101
seminiferous tubules 100
septal cells 63
serotonin 2
Sertoli cells 105
serum 14
sex hormones 2
sinoatrial node 27
sinusoids 95
skin 53
small intestine 90
sodium 79, 80, 81

somatocrinin 4
somatomedins 4
somatostatin 4, 9
somatotrophs 4
somatotropin 4
specific defense mechanisms 55
sperm 9, 98, 104, 108
spermatic cord 100
spermatids 104
spermatogenesis 104, 105
spermatogenic cells 104
spermatogonia 104
spermatozoa 104
spermiation 105
spermiogenesis 104, 105
sphincter of Oddi 95
sphygmomanometer 38
spleen 51, 52
    red pulp 52
    white pulp 52
stem cells 16, 19
    hemocytoblast 19
sternocleidomastoid muscle 65
steroid hormones 1, 2
stomach 89
stroke 21
sublingual gland 88
submandibular gland 88
submucosa 86
superior vena cava 23, 43
surfactant 63
suspensory ligament 102
sustentacular cells 104, 105
systole 28

T

T-cells 51, 52, 56
    activated 59
    helper 59
    killer 59
    memory 59
    suppressor 59
T4 cells 56
T8 cells 56
tachycardia 38
taeniae coli 96
taste buds 87
testes 9, 11, 98, 99, 100
testosterone 2, 11, 105
tetraiodothyronine 6
thecal cells 106
thoracic duct 51

throat 62
thrombin 20
thrombocytes 19
thymosin 11, 52
thymus 11, 51
thymus gland 51, 52
thyroid gland 6
thyroid hormones 2, 6
thyroid-stimulating hormone 4, 6
thyrotrophs 4
thyrotropin releasing hormone 6
tidal volume 68
tongue 87
tonsils 51, 52
   lingual 52, 62
   palatine 52, 62
   pharyngeal 52, 62
total lung capacity 68
trabeculae carnae 24
trachea 63
transferrin 16, 53
trigone 83
triiodothyronine 6
trimester
   first 110
   second 110
   third 110
trophoblast 108
trypsin 96
tubular secretion 80
tubulus rectus 100
tunica adventitia 32
tunica albuginea 100, 102
tunica externa 32
tunica interna 32, 33
tunica media 32, 33
tunica vaginalis 99

U

umbilical cord 110
umbilicus 48
universal donors 17
universal recipients 17
urea 78, 82
ureters 71, 73, 82
urethra 71, 100
   prostatic 100
   spongy 100, 101
urethral orifice
   external 83, 100, 101
   internal 83
uric acid 78, 82

urinalysis 81
urinary bladder 71, 82
urinary system 71
urination 83
urine 74, 77, 81, 82
urine pH 82
urochrome 81
urogastrone 90
urogenital diaphragm 100
uterine wall 103
uterosacral ligaments 103
uterus 103, 110
uvula 86

V

vagina 103, 108, 110
vaginal vestibule 104
vas deferens 100
vasa recta 77
vasa vasorum 32
vasoconstriction 37
vasodilation 37
vasomotor center 37
vasopressin 4
veins 33
   accessory azygos 46
   anterior tibial 46
   arcuate 77
   axillary 45
   azygos 46
   basilic 45
   brachiocephalic 43
   cephalic 45
   common iliac 46
   external iliac 46
   external jugular 43
   femoral 46
   gastric 48
   great saphenous 46
   hemiazygos 46
   hepatic 46, 48
   hepatic portal 48
   inferior mesenteric 48
   interlobar 77
   interlobular 77
   internal iliac 46
   internal jugular 43
   major superficial 45
   median antebrachial 45
   median cubital 45
   peroneal 46
   plantar 46

popliteal 47
posterior tibial 46
pulmonary 31
segmental 77
small saphenous 46
splenic 48
subclavian 43
superior mesenteric 48
systemic 31
umbilical 48
vertebral 43
ventricles 23
venules 32
vermiform appendix 52, 96
vestibule 86
villi 91
vital capacity 68
vocal cords 62

vocal folds 62
voice box 62
vulva 104

W

white blood cells 19
whole blood 12
windpipe 63

Y

yolk sac 110

Z

zona glomerulosa 7
zona pellucida 106, 108
zygote 108